D1649198

Volume One

Student Solutions Manual for Reese's
University Physics

Ronald Lane Reese
Washington & Lee University

Robin B. S. Brooks
Bates College

Mark D. Semon
Bates College

Brooks/Cole Publishing Company

I(T)P® An International Thomson Publishing Company

Pacific Grove • Albany • Belmont • Bonn • Boston • Cincinnati • Detroit • Johannesburg • London
Madrid • Melbourne • Mexico City • New York • Paris • Singapore • Tokyo • Toronto • Washington

Sponsoring Editor: *Melissa D. Henderson*
Editorial Assistant: *Dena Dowsett-Jones*
Marketing Manager: *Steve Catalano*
Marketing Assistant: *Christina De Veto*

Production Coordinator: *Dorothy Bell*
Cover Design: *Roy R. Neuhaus*
Cover Photo: *Ronald Lane Reese*
Printing and Binding: *Patterson Printing*

For more information, contact:

BROOKS/COLE PUBLISHING COMPANY
511 Forest Lodge Road
Pacific Grove, CA 93950
USA

International Thomson Editores
Seneca 53
Col. Polanco
11560 México, D. F., México

International Thomson Publishing Europe
Berkshire House 168-173
High Holborn
London WC1V 7AA
England

International Thomson Publishing GmbH
Königswinterer Strasse 418
53227 Bonn
Germany

Thomas Nelson Australia
102 Dodds Street
South Melbourne, 3205
Victoria, Australia

International Thomson Publishing Asia
60 Albert Street
#15-01 Albert Complex
Singapore 189969

Nelson Canada
1120 Birchmount Road
Scarborough, Ontario
Canada M1K 5G4

International Thomson Publishing Japan
Palaceside Building, 5F
1-1-1 Hitotsubashi
Chiyoda-ku, Tokyo 100-0003
Japan

Printed in the United States of America

10   9   8   7   6   5   4   3   2   1

ISBN 0-534-35234-0

Contents

To Our Students

This solutions manual is for you. One of the best ways to learn is through apprenticeship; however, since we can't work side by side with you, we offer this book in our place. We hope reading it will be like looking over our shoulders as we solve the problems, and that you will benefit from seeing how we, as more experienced problem solvers, proceed.

Volume 1 of the *Student Solutions Manual* contains solutions to every other odd problem (1, 5, 9, 13, ...) in Chapters 1 through 15 of *University Physics*; Volume 2 contains solutions to every other odd problem in Chapters 13 through 27. For your convenience, problems in *University Physics* are numbered in red if their solutions are in either Volume 1 or Volume 2 of the *Student Solutions Manual*.

More often than not, there are several different ways of solving a problem (see page 12 of *University Physics* for a discussion of this), so it's all right if your way differs from ours – as long as your method is correct! The most beneficial way of using this book is to attempt the problem yourself first, and only then to look at how we solve it. It is one thing to solve a problem yourself and a totally different thing to read someone else's solution. If you read our solution first, hopefully you will think: this makes sense, that makes sense, and so on; but then, chances are you will not be able to solve the problem when you encounter it on a test or in some other situation in which you can't refer to the solutions manual. We don't think you want to find yourself in that position! On the other hand, if you try the problem *before* reading our solution and get stuck, you create in yourself a "need to know" which makes you more likely to truly grasp the point(s) you didn't fully understand.

Physics is not a spectator sport. Just as listening to a piano concerto is not the same as playing it yourself, watching (or reading) how someone else solves problems won't improve *your* ability to handle them. Piano playing and problem solving both take consistent practice to master.

Although the problems in *University Physics* span a wide variety of subjects, there are some common elements in how we suggest you approach them. (See page 12 of *University Physics* for an elaboration on this.) First, whenever possible, make a sketch of the situation involved. Second, identify which quantities you know and which you wish to know. Third, review the laws and equations involving these quantities to see how what you wish to know is related to what you already know. Also, pay attention to the Problem – solving Tactics discussed in each chapter of the text; they are there to help you solve different types of problems and avoid common pitfalls along the way. All this may sound simple, but most of us need a lot of practice to really learn it, and to develop the conviction that the answers we seek really are contained in the laws we have learned – no matter how unlikely this may seem at first glance!

In some cases your final numerical solution may differ slightly from ours. This is most likely due to the number of 'significant digits' you carry through your calculations. We have expressed our answers to the precision permitted by the numbers upon which they are based, so if your answer differs slightly from ours, be sure to check whether you are using the proper number of significant figures. (See pages 18-21 of *University Physics* for a more complete discussion of this.) In some cases, however, an answer will depend upon where the correct number of significant digits is imposed. For example, in problems with several parts, we quote the answer to each part with the proper number of significant digits and calculate what follows using this number. Because of this, answers in the parts that follow might differ from what you would have found if you hadn't stopped along the way to quote an intermediate result. If you prefer to complete the whole problem before imposing the correct number of significant figures, don't worry if your answer differs slightly from ours.

Also, different people have different 'rounding' conventions. For example, in a situation where only two significant digits are allowed, some people will round a number like 7.65 up to 7.7 while others will round it down to 7.6. We have chosen the former convention, and in fact, we round all numbers *up* whose first 'insignificant digit' is a five, regardless of whether the digit preceding it is even or odd.

We have made an effort to keep our notation and terminology consistent with the text. There may, however be a few small differences, in particular we use a bold face zero **0** to denote the zero vector, while the text uses the same notation, 0, for both the scalar and the zero vector.

One of the most common statements a teacher hears is "I understand the material but I just can't do the problems." Although problem-solving is a skill in its own right, we believe that if you cannot work the problems you do not fully understand the material. You might *know* the material, but you have not really *understood* it. In our view, *understanding* is the result of *knowledge and experience*, and *experience* only comes from *using* the material in hands-on activities, like solving problems and doing laboratory experiments.

Finally, we want to encourage you heartily! Best wishes for success on your journey into physics, engineering, mathematics, and other sciences. It is not an easy road, but it is one we have found to be intellectually rewarding and enriching. Some people (erroneously) believe that learning science is memorizing a large collection of facts and formulas. We hope you will see how miraculous it is that most of these facts and formulas can be understood within the framework of remarkably few laws of nature. We also hope that in solving these problems, you experience the bewilderment of seeing something you have not seen before, the thrill of understanding it, and the wonder of seeing how it all fits together into a bigger picture.

We had a good time doing these problems, and we hope you do too!

Ronald Lane Reese   Robin B.S. Brooks   Mark D. Semon
Dept of Physics   Dept of Mathematics   Dept of Physics
Washington and Lee University Bates College   Bates College
Lexington, VA 24450   Lewiston, ME 04240   Lewiston, ME 04240
reeser@wlu.edu   rbrooks@bates.edu   msemon@bates.edu

January 25, 1999

Acknowledgments

We thank our editors at Brooks/Cole for all their help. In particular, we thank Beth Wilbur for her encouragement, support, and dedication to excellence.

We also thank our "Checkers," those anonymous teachers who discovered typographical errors in what we *thought* was perfect copy.

Our *Student Solutions Manual* was typeset entirely on a personal computer using LaTeX, an extension of TeX. The TeXprogram and language were created by Donald Knuth. We thank Professor Knuth and the many people following him for their marvelous gift to the to the world.

Chapter 1

Preludes

1.1 The volume V of a cylinder is the product of the area of the base, πr^2, and the height h:

$$V = \pi r^2 h.$$

Using the dimensions given for the standard kilogram at 0 °C:

$$V = \pi \left(\frac{3.8948}{2}\ \text{cm}\right)^2 3.8948\ \text{cm} = 46.403\ \text{cm}^3 \left(\frac{1\ \text{m}}{100\ \text{cm}}\right)^3 = 4.6403 \times 10^{-5}\ \text{m}^3,$$

neglecting the slight beveling of the edges. Note that this is about 46 times the volume of one gram of water.

1.5 Reading across —

peekaboos	nanny goats	military	millionaires
megaphones	microphones	megabucks	centipedes
microfilm	terrapin	megalopolis	deck of cards
microfiche	microwaves	decimate	microeconomics
to kill a mockingbird	terrible	terra firma	megalomaniacs
millinery	microscopes	megabytes	

1.9

a) $\dfrac{\text{mass of a proton}}{\text{mass of an electron}} = \dfrac{1.67 \times 10^{-27}\ \text{kg}}{9.11 \times 10^{-31}\ \text{kg}} = 1.83 \times 10^3.$

This ratio occurs frequently in Physics, and usually is expressed to four significant figures as 1836 (the year of the Alamo!). The number 1836 is obtained by using the "more precise values" from the inside cover of the text and then keeping only four significant figures.

b) $\dfrac{\text{mass of the Sun}}{\text{mass of the Earth}} = \dfrac{1.99 \times 10^{30}\ \text{kg}}{5.98 \times 10^{24}\ \text{kg}} = 3.33 \times 10^5.$

c) $\dfrac{\text{mass of the Galaxy}}{\text{mass of the Sun}} \approx \dfrac{3 \times 10^{41}\ \text{kg}}{1.99 \times 10^{30}\ \text{kg}} = 2 \times 10^{11}.$ (to one sig. fig.)

d) $\dfrac{\text{mass of the Universe}}{\text{mass of the Milky Way Galaxy}} \approx \dfrac{1 \times 10^{53}\ \text{kg}}{3 \times 10^{41}\ \text{kg}} = 3 \times 10^{11}.$

Note that the ratios in c) and d) have the same order of magnitude.

1.13

a)   From Table 1.5, the most distant objects in the visible universe are about 10^{26} m away, so a length a factor of 10^{20} smaller than this is about

$$\frac{10^{26} \text{ m}}{10^{20}} = 10^6 \text{ m}.$$

b)   From Table 1.5, a length a factor of 10^{20} greater than the diameter of the nucleus of an atom is $10^{20} \times 10^{-15}$ m $= 10^5$ m.

1.17

a)   Let ℓ be the length and w the width. The original perimeter is $P = 2\ell + 2w$, and the new perimeter is $P_{\text{new}} = 2(3\ell) + 2(3w) = 3(2\ell + 2w)$. Therefore, the perimeter increases by a factor of **3**.

b)   The original area is $A = \ell w$, and the new area is $A_{\text{new}} = (3\ell)(3w) = 9\ell w$. Therefore, the area increases by a factor of nine.

1.21

a)   The surface area of a sphere is $4\pi r^2$. If the surface area triples, then $4\pi r^2_{\text{new}} = 3(4\pi r^2)$. Therefore, the ratio of the radii is $\sqrt{3} = 1.73$, so the radius increases by a factor of $\sqrt{3}$.

b)   The volume of a sphere is $(4/3)\pi r^3$. Therefore, the increase in volume is the cube of the increase in radius, a factor of $(\sqrt{3})^3 = 3\sqrt{3} = 5.20$.

1.25   There are 5280 feet in one mile. Each foot is 12 inches, and each inch is 2.54 centimeters. Hence, the number of centimeters in one mile is

$$\left(5280 \frac{\text{ft}}{\text{mile}}\right) \left(12 \frac{\text{in}}{\text{ft}}\right) \left(2.54 \frac{\text{cm}}{\text{in}}\right) = 1.61 \times 10^5 \frac{\text{cm}}{\text{mile}} = 1.61 \times 10^3 \frac{\text{m}}{\text{mile}}.$$

Let v be the numerical value of the speed in miles per hour. To convert this to (m/s):

$$v \text{ in m/s} = \left(v \frac{\text{mile}}{\text{h}}\right) \left(1.61 \times 10^3 \frac{\text{m}}{\text{mile}}\right) \left(\frac{\text{h}}{3.600 \times 10^3 \text{ s}}\right) = v \times 0.447 \text{ m/s}.$$

Therefore, to convert the numerical value of v in miles per hour to a numerical value in meters per second, multiply the numerical value of v in miles per hour by the factor 0.447.

1.29   The density of water is

$$1.0 \times 10^3 \text{ kg/m}^3 = 1.0 \text{ g/cm}^3.$$

Hence, the number of moles in one cubic centimeter of water, i.e., 1.0 g of water, is

$$\frac{1.0 \text{ g}}{18 \text{ g/mol}} = 0.056 \text{ mol}.$$

Each mole has Avogadro's number of particles. Therefore, the number of molecules in the cubic centimeter of water is

$$0.056 \text{ mol} = 0.056 \text{ mol} \left(6.022 \times 10^{23} \frac{\text{particle}}{\text{mol}}\right) = 3.4 \times 10^{22} \text{ particle}.$$

1.33   There are twelve inches per foot and 2.54 cm/in. Hence, the length of each edge of a cord of wood is

$$4.00 \text{ ft} \left(12 \frac{\text{in}}{\text{ft}}\right) \left(2.54 \frac{\text{cm}}{\text{in}}\right) \left(10^2 \frac{\text{m}}{\text{cm}}\right) = 1.22 \text{ m}, \quad \text{and}$$

$$8.00 \text{ ft} \left(12 \frac{\text{in}}{\text{ft}}\right) \left(2.54 \frac{\text{cm}}{\text{in}}\right) \left(10^2 \frac{\text{m}}{\text{cm}}\right) = 2.44 \text{ m}.$$

Therefore, the metric volume of a cord of wood is

$$(1.22 \text{ m})(1.22 \text{ m})(2.44 \text{ m}) = 3.63 \text{ m}^3.$$

1.37

a) The mass m of the brick is the product of its density ρ with its volume V, $m = \rho V$. Therefore, its volume is $V = \dfrac{m}{\rho} = \dfrac{15.0 \text{ kg}}{19.3 \times 10^3 \text{ kg/m}^3} = 7.77 \times 10^{-4} \text{ m}^3.$

b) The increase in the volume of the brick is the volume of the 3.0 kg of additional gold: $V = \dfrac{m}{\rho} = \dfrac{3.0 \text{ kg}}{19.3 \times 10^3 \text{ kg/m}^3} = 1.6 \times 10^{-4} \text{ m}^3.$

1.41

a) The volume of a single person is $[0.200 \text{ m}] [0.300 \text{ m}] [1.80 \text{ m}] = 0.108 \text{ m}^3$. Therefore, 6.0×10^9 people have a total volume of $[6.0 \times 10^9 \text{ person}] [0.108 \text{ m}^3/\text{person}] = 6.5 \times 10^8 \text{ m}^3.$

b) Let ℓ be the length of the edge of a cube capable of containing the entire human population. Then $\ell^3 = 6.5 \times 10^8 \text{ m}^3$, so $\ell = 8.7 \times 10^2 \text{ m}.$

c) The area occupied by a single person is $[0.200 \text{ m}] [0.300 \text{ m}] = 0.0600 \text{ m}^2$. The area occupied by the entire human population, then, is $[6.0 \times 10^9 \text{ person}] [0.0600 \text{ m}^2/\text{person}] = 3.6 \times 10^8 \text{ m}^2.$

d) Let ℓ be the length of the edge of a square that contains the entire human population. Then $\ell^2 = 3.6 \times 10^8 \text{ m}^2$, so $\ell = 19 \times 10^3 \text{ m}.$

1.45 The conversion of the speed of light from (m/s) to (furlongs/fortnight) is as follows: The number of seconds per fortnight is

$$[14 \text{ d/fortnight}] [24 \text{ h/d}] [60 \text{ min/h}] [60 \text{ s/min}] = 1.21 \times 10^6 \text{ s/fortnight}.$$

The number of meters per furlong is

$$[0.125 \text{ mile/furlong}] [1.61 \times 10^3 \text{ m/mile}] = 2.01 \times 10^2 \text{ m/furlong}.$$

Therefore, the speed of light is

$$\left[\frac{2.9979 \times 10^8 \text{ m/s}}{2.01 \times 10^2 \text{ m/furlong}} \right] [1.21 \times 10^6 \text{ s/fortnight}] = 1.80 \times 10^{12} \text{ furlongs/fortnight}.$$

1.49

a) Since the mass of 100 freshly minted \$1 bills is about 0.0974 kg, the mass of a single bill is 9.74×10^{-4} kg.

b) Since the thickness of a stack of 100 freshly minted \$1 bills is about 8.0×10^{-3} m, an individual bill is approximately 8.0×10^{-5} m thick.

c) Since the dimensions of a dollar bill are: length $= 0.156$ m, width $= 0.066$ m, thickness $= 8.0 \times 10^{-5}$ m, the volume is $8.2 \times 10^{-7} \text{ m}^3.$

d) Use the result of part a). The mass of 10^9 dollar bills is: 10^9 bill $\times 9.74 \times 10^{-4}$ kg/bill $= 9.74 \times 10^5$ kg.

e) Use the result of part b). The height of 10^9 dollar bills is: 10^9 bill $\times 8.0 \times 10^{-5}$ m/bill $= 8.0 \times 10^4$ m.

f) The area of 10^9 dollar bills is $10^9 \times 1.56 \times 10^{-1}$ m $\times 6.6 \times 10^{-2}$ m $= 1.0 \times 10^7 \text{ m}^2.$

g) The volume of 10^9 dollar bills is 10^9 bill $\times 8.2 \times 10^{-7} \text{ m}^3/\text{bill} = 8.2 \times 10^2 \text{ m}^3$. Let ℓ be the length of the side of cube with the same volume as 10^9 dollar bills. Then $\ell^3 = 8.2 \times 10^2$, so $\ell = 9.4 \text{ m}.$

1.53

a)   To convert the density to kilograms per cubic meter:

$$5 \times 10^{-31} \text{ g/cm}^3 = 5 \times 10^{-31} \text{ g/cm}^3 \left(\frac{1 \text{ kg}}{10^3 \text{ g}}\right) \left(\frac{100 \text{ cm}}{1 \text{ m}}\right)^3 = 5 \times 10^{-28} \text{ kg/m}^3.$$

b)   To see how much of the universe belongs to each of us, divide our mass by the volume of a sphere and set it equal to our last answer. Taking our mass to be 60 kg, we have

$$\frac{60 \text{ kg}}{4\pi r^3/3} = 5 \times 10^{-28} \text{ kg/m}^3.$$

Solving, we find $r = 3 \times 10^9$ m. Thus each of us is entitled to a sphere with a radius about 2000 times that of the Earth!

1.57

a)   A good swimmer has a speed about equal to that of a person walking—about 5 km/h. So the time necessary to swim 1 km is about $(1 \text{ km})/(5 \text{ km/h})$, or about 12 min.

b)   The circumference of the Earth is about 4.0×10^4 km. Perhaps an individual can paddle about 40 kilometers a day under the best of conditions. So the shortest time would be about 10^3 d or 3 y.

c)   As an exhausted student or professor, you probably nod off in several minutes, so about 10 min.

1.61   To express the mass of the Sun in units of the proton mass, set

$$1.99 \times 10^{30} \text{ kg} = 1.99 \times 10^{30} \text{ kg} \left(\frac{1 \text{ proton mass}}{1.67 \times 10^{-27} \text{ kg}}\right) = 1.19 \times 10^{57} (\text{ proton mass}).$$

1.65

a)   Assume about 3 ties are needed per meter of each track or 6 tie/m for both tracks. The distance between New York and San Francisco is about 5×10^3 km $= 5 \times 10^6$ m. Hence, the total number of ties needed is $[6 \text{ tie/m}][5 \times 10^6 \text{ m}] \approx 3 \times 10^7$ tie, or about 30 million ties!

b)   The dimensions of a tie are about 15 cm by 15 cm by 2.0 m, so the volume of a single tie is about $[0.15 \text{ m}][0.15 \text{ m}][2.0 \text{ m}] \approx 0.045 \text{ m}^3$. Therefore, the total volume of concrete needed to make the ties is $[3 \times 10^7 \text{ tie}][0.045 \text{ m}^3/\text{tie}] \approx 1 \times 10^6 \text{ m}^3$.

1.69   The radius of the Earth is 6.37×10^3 km. The surface area A of the spherical Earth is $A = 4\pi r^2 = 4\pi[6.37 \times 10^3 \text{ km}]^2 = 5.10 \times 10^8 \text{ km}^2$.

   About 70% of the surface of the Earth is covered with oceans, so the surface area of the oceans is about $0.70A = [0.70][5.10 \times 10^8 \text{ km}^2] = 3.6 \times 10^8 \text{ km}^2$.

   Assume the average depth of the ocean is about 2 km. Then the volume of the oceans is about $[2 \text{ km}][3.6 \times 10^8 \text{ km}^2] = 7 \times 10^8 \text{ km}^3$. Note that since the density of seawater is $1.025 \times 10^3 \text{ kg/m}^3$, the total mass of the oceans is about 7×10^{20} kg.

1.73   Assume that McDonalds has been in existence for about 30 years. That means an average yearly consumption of about $\dfrac{100 \times 10^9 \text{ burger}}{30 \text{ y}} \approx 3.0 \times 10^9$ burger/y.

   However, during the early years of its existence, its output was probably much less, while the current output likely is more. Hence, take the yearly output to be about 5.0×10^9 burger/y. Then the daily consumption of burgers, is about $\dfrac{5.0 \times 10^9 \text{ burger/y}}{365 \text{ d/y}} = 1.4 \times 10^7$ burger/d.

   Assume each burger is bought by a different person. Then 1.4×10^7 people must buy a McDonalds burger each day. The population of the United States is about 2.5×10^8 people. Hence, the fraction of the population that visits McDonalds each day for a burger is about $\dfrac{1.4 \times 10^7 \text{ person}}{2.5 \times 10^8 \text{ person}} = 5.6 \times 10^{-2} \approx 6\%$.

1.77   For each angle, one can keep 3, 4, and 5 significant figures, respectively.   The results are tabulated below.

angle	sine	cosine	tangent
78.0°	0.978	0.208	4.70
78.02°	0.9782	0.2076	4.713
78.024°	0.97823	0.20750	4.7143

The cosine of a small angle is nearly 1, so cosine cannot be included in the small angle approximations.

Chapter 2

A Mathematical Tool Box

2.1 Since any two copies of a vector have the same length and also are parallel, make two copies of each of the vectors $\vec{F}_1$ and $\vec{F}_2$, and arrange them so they form the sides of a *parallelogram* as shown below:

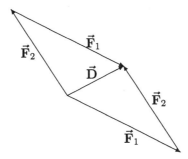

In the top triangle of the parallelogram, we have $\vec{F}_2$ followed by $\vec{F}_1$. Therefore, the diagonal vector $\vec{D}$ must be $\vec{F}_2 + \vec{F}_1$.

On the other hand, the bottom triangle has the vector $\vec{F}_1$ followed by $\vec{F}_2$, so the diagonal vector $\vec{D}$ must be $\vec{F}_1 + \vec{F}_2$.

Therefore $\vec{F}_2 + \vec{F}_1 = \vec{D} = \vec{F}_1 + \vec{F}_2$.

2.5 The stroll is sketched below.

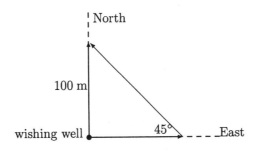

Since north and east are perpendicular to each other, the two strolls and their vector sum form a $45°$ right triangle. Such a triangle has two equal sides. Hence, the vector to the east has magnitude $100\,\text{m}$. The vector to the northwest is the hypotenuse of the right triangle, and so has magnitude $\sqrt{(100\,\text{m})^2 + (100\,\text{m})^2} = 141\,\text{m}$.

2.9   Two vectors are perpendicular to each other if and only if their scalar product is zero. Hence to find out if $\vec{R} + \vec{r}$ is perpendicular to $\vec{R} - \vec{r}$, compute their scalar product:

$$(\vec{R} + \vec{r}) \bullet (\vec{R} - \vec{r}) = \vec{R} \bullet \vec{R} - \vec{R} \bullet \vec{r} + \vec{r} \bullet \vec{R} - \vec{r} \bullet \vec{r}$$

$$= \vec{R} \bullet \vec{R} - \vec{r} \bullet \vec{r} \qquad \text{Since } \vec{R} \bullet \vec{r} = \vec{r} \bullet \vec{R}.$$

$$= R^2 - r^2$$

$$= 0 \qquad \text{(Since the two vectors have the same magnitude.)}$$

The zero scalar product implies that the two vectors are perpendicular.

It makes no difference which way the vector difference is taken; that is, $\vec{R} + \vec{r}$ also is perpendicular to $\vec{r} - \vec{R}$, since $\vec{R} - \vec{r}$ and $\vec{r} - \vec{R}$ are antiparallel.

2.13   The scalar product of two vectors is defined as

$$\vec{A} \bullet \vec{B} = AB \cos\theta,$$

where θ is the angle between them when their tails are at a common point.

If $\vec{A}$ and $\vec{B}$ are the *same* vector, then $\theta = 0°$, $\cos\theta = 1$, and the dot product is just the square of the magnitude, which in the case of unit vectors is 1. Thus

$$\hat{i} \bullet \hat{i} = \hat{j} \bullet \hat{j} = \hat{k} \bullet \hat{k} = 1.$$

If $\vec{A}$ and $\vec{B}$ are *different*, but each is one of the standard unit vectors $\hat{i}$, $\hat{j}$, or $\hat{k}$, then $\theta = 90°$, $\cos\theta = 0$, and the dot product is 0. Thus

$$\hat{i} \bullet \hat{j} = \hat{i} \bullet \hat{k} = \hat{j} \bullet \hat{i} = \hat{j} \bullet \hat{k} = \hat{k} \bullet \hat{i} = \hat{k} \bullet \hat{j} = 0.$$

2.17   The magnitude of the vector $\hat{i} + \hat{j} + \hat{k}$ is $\sqrt{(1)^2 + (1)^2 + (1)^2} = \sqrt{3}$. If you divide a vector by its magnitude, the resulting vector is a vector one unit long in the same direction as the original. Therefore,

$$\frac{\hat{i} + \hat{j} + \hat{k}}{\sqrt{3}} = \frac{1}{\sqrt{3}}\hat{i} + \frac{1}{\sqrt{3}}\hat{i} + \frac{1}{\sqrt{3}}\hat{i}$$

is in the same direction as $\hat{i} + \hat{j} + \hat{k}$ but is exactly one unit long.

To check this, compute

$$\left| \frac{1}{\sqrt{3}}\hat{i} + \frac{1}{\sqrt{3}}\hat{i} + \frac{1}{\sqrt{3}}\hat{i} \right| = \sqrt{\left(\frac{1}{\sqrt{3}}\right)^2 + \left(\frac{1}{\sqrt{3}}\right)^2 + \left(\frac{1}{\sqrt{3}}\right)^2} = \sqrt{\frac{3}{3}} = 1.$$

2.21   Here is a sketch of the relationship among the vectors.

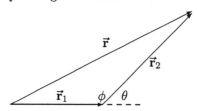

Notice that θ is the angle between $\vec{r}_1$ and $\vec{r}_2$ when their tails are put at a common point. Therefore $\vec{r}_1 \bullet \vec{r}_2 = r_1 r_2 \cos\theta$. Use this to compute

$$r^2 = (\vec{r}_1 + \vec{r}_2) \bullet (\vec{r}_1 + \vec{r}_2) = r_1^2 + \vec{r}_1 \bullet \vec{r}_2 + \vec{r}_2 \bullet \vec{r}_1 + r_2^2 = r_1^2 + 2r_1 r_2 \cos\theta + r_2^2.$$

This is almost, but not quite, the "law of cosines." The law of cosines is generally stated in terms of the interior angle ϕ rather than the exterior angle θ. Since $\theta + \phi = 180°$, then $\cos\theta = -\cos\phi$. Hence, the above equation implies

$$r^2 = r_1^2 + r_2^2 - 2r_1 r_2 \cos\phi,$$

which is the law of cosines.

2.25

a)   For any two non-zero vectors $\vec{\mathbf{A}}$ and $\vec{\mathbf{B}}$,

$$\vec{\mathbf{A}} \bullet \vec{\mathbf{B}} = AB\cos\theta \implies \cos\theta = \frac{\vec{\mathbf{A}} \bullet \vec{\mathbf{B}}}{AB},$$

where θ is the angle between them when their tails are at a common point. We'll use this last equation to find the angle θ between $\vec{\mathbf{A}} = 3.00\hat{\mathbf{i}} + 4.00\hat{\mathbf{j}} + 5.00\hat{\mathbf{k}}$ and $\vec{\mathbf{B}} = \hat{\mathbf{i}}$. First find $\vec{\mathbf{A}} \bullet \vec{\mathbf{B}}$:

$$\vec{\mathbf{A}} \bullet \vec{\mathbf{B}} = (3.00\hat{\mathbf{i}} + 4.00\hat{\mathbf{j}} + 5.00\hat{\mathbf{k}}) \bullet \hat{\mathbf{i}} = (3.00)(1.00) + (4.00)(0.00) + (5.00)(0.00) = 3.00.$$

Now find $A = |\vec{\mathbf{A}}|$ and $B = |\vec{\mathbf{B}}|$:

$$A = |3.00\hat{\mathbf{i}} + 4.00\hat{\mathbf{j}} + 5.00\hat{\mathbf{k}}| = \sqrt{(3.00)^2 + (4.00)^2 + (5.00)^2} = 7.07 \qquad \text{and} \qquad B = |\hat{\mathbf{i}}| = 1.$$

Thus

$$\cos\theta = \frac{3.00}{(7.07)(1.00)} = 0.424 \implies \theta = 64.9°.$$

b)   The angle between $\hat{\mathbf{i}}$ and $-\hat{\mathbf{i}}$ is $180°$. So the angle that $\vec{\mathbf{A}}$ makes with $-\hat{\mathbf{i}}$ is $180.0° - 64.9° = 115.1°$. This answer also may be found by using the methods and results of part a):

$$\frac{\vec{\mathbf{A}} \bullet (-\hat{\mathbf{i}})}{A| - \hat{\mathbf{i}}|} = \frac{-\vec{\mathbf{A}} \bullet \hat{\mathbf{i}}}{A|\hat{\mathbf{i}}|} = -0.424 = -\cos 64.9° = +\cos(180.0° - 64.9°) = \cos 115.1°,$$

so the angle between $\vec{\mathbf{A}}$ and $-\hat{\mathbf{i}}$ is $115.1°$.

2.29

a)   The magnitudes of $\vec{\mathbf{A}} = \cos\alpha\,\hat{\mathbf{i}} + \sin\alpha\,\hat{\mathbf{j}}$ and $\vec{\mathbf{B}} = \cos\beta\,\hat{\mathbf{i}} + \sin\beta\,\hat{\mathbf{j}}$ are

$$A = \sqrt{(\cos\alpha)^2 + (\sin\alpha)^2} = \sqrt{1} = 1 \qquad \text{and} \qquad B = \sqrt{(\cos\beta)^2 + (\sin\beta)^2} = \sqrt{1} = 1.$$

b)   Both vectors lie in the x–y plane. Here is a sketch of the vectors $\vec{\mathbf{A}}$, $\vec{\mathbf{B}}$, and $\hat{\mathbf{i}}$. We've assumed $\alpha > \beta$.

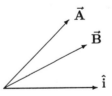

Since these are all unit vectors, the scalar product of any two of them is the cosine of the angle between them. Since $\vec{\mathbf{A}} \bullet \hat{\mathbf{i}} = (\cos\alpha\,\hat{\mathbf{i}} + \sin\alpha\,\hat{\mathbf{j}}) \bullet \hat{\mathbf{i}} = \cos\alpha$, the angle from $\hat{\mathbf{i}}$ to $\vec{\mathbf{A}}$ must be α. Similarly, the angle from $\hat{\mathbf{i}}$ to $\vec{\mathbf{B}}$ is β. Hence, from the picture, the angle from $\vec{\mathbf{B}}$ to $\vec{\mathbf{A}}$ must be $\alpha - \beta$. Therefore,

$$\cos(\alpha - \beta) = \vec{\mathbf{A}} \bullet \vec{\mathbf{B}} = (\cos\alpha\,\hat{\mathbf{i}} + \sin\alpha\,\hat{\mathbf{j}}) \bullet (\cos\beta\,\hat{\mathbf{i}} + \sin\beta\,\hat{\mathbf{j}}) = \cos\alpha\cos\beta + \sin\alpha\sin\beta.$$

2.33

a)   Let $\vec{\mathbf{A}}$ be the vector with magnitude 12, $\vec{\mathbf{B}}$ be the vector with magnitude 16, and $\vec{\mathbf{C}} = \vec{\mathbf{A}} + \vec{\mathbf{B}}$. Then the facts of the problem are

$$\vec{\mathbf{C}} = \vec{\mathbf{A}} + \vec{\mathbf{B}}, \quad A = 12, \quad B = 16, \quad \text{and} \quad C = 20.$$

We want to find the angle between $\vec{A}$ and $\vec{B}$. To do this, we need to compute $\vec{A} \bullet \vec{B}$ in order to find the cosine of the angle between them. Since we know the magnitudes A, B, and C, the trick is to find a relation among these three magnitudes and the scalar product $\vec{A} \bullet \vec{B}$. This is accomplished by computing $\vec{C} \bullet \vec{C}$:

$$C^2 = \vec{C} \bullet \vec{C} = (\vec{A} + \vec{B}) \bullet (\vec{A} + \vec{B}) = \vec{A} \bullet \vec{A} + \vec{A} \bullet \vec{B} + \vec{B} \bullet \vec{A} + \vec{B} \bullet \vec{B} = A^2 + 2\vec{A} \bullet \vec{B} + B^2.$$

Therefore,

$$20^2 = 12^2 + 2\vec{A} \bullet \vec{B} + 16^2 \implies \vec{A} \bullet \vec{B} = \frac{1}{2}(20^2 - 12^2 - 16^2) = 0.$$

Since $\vec{A} \bullet \vec{B} = 0$, the two vectors are perpendicular, i.e., the angle between them is $90°$.

The three vectors form a right triangle as depicted below.

This is the familiar right triangle whose sides are in the proportions 3, 4, and 5.

b) Let α be the angle between $\vec{A}$ and $\vec{C}$, and let β be the angle between $\vec{B}$ and $\vec{C}$. To find the angle α use the result from part a) to compute $\cos \alpha$:

$$\vec{A} \bullet \vec{C} = \vec{A} \bullet (\vec{A} + \vec{B}) = \vec{A} \bullet \vec{A} + \vec{A} \bullet \vec{B} = A^2 + 0 = 144.$$

Therefore

$$\cos \alpha = \frac{\vec{A} \bullet \vec{C}}{AC} = \frac{144}{(12)(20)} = 0.60 \implies \alpha = 53°.$$

Similarly

$$\vec{B} \bullet \vec{C} = \vec{B} \bullet (\vec{A} + \vec{B}) = \vec{B} \bullet \vec{B} + \vec{A} \bullet \vec{B} = B^2 + 0 = 256.$$

So,

$$\cos \beta = \frac{\vec{B} \bullet \vec{C}}{BC} = \frac{256}{(16)(20)} = 0.80 \implies \beta = 37°.$$

These results also could have been obtained using elementary trigonometry on the right triangle depicted in part a).

2.37 Two vectors are equal to each other if and only if their respective components are equal to each other. Hence

$$3.00 = 2.00 - \alpha \implies \alpha = -1.00$$
$$4.00 = -(6.00 + \beta) \implies \beta = -10.00$$
$$-6.00 = 3.00 - \gamma \implies \gamma = 9.00.$$

2.41 If you draw a picture of $\hat{i} + \hat{j}$ in the ordinary x-y plane, you'll see that it goes up and to the right at a $45°$ angle to the x-axis. This suggests choosing a vector of equal length that goes up and to the *left* at a $45°$ angle to the x-axis as the required vector perpendicular to $\hat{i} + \hat{j}$. This vector is $-\hat{i} + \hat{j}$. In order to check algebraically that $-\hat{i} + \hat{j}$ is perpendicular to $\hat{i} + \hat{j}$, compute their scalar product:

$$(-\hat{i} + \hat{j}) \bullet (\hat{i} + \hat{j}) = -1 + 1 = 0.$$

Since the scalar product is zero, the two vectors are, in fact, perpendicular. Any scalar multiple of $-\hat{i} + \hat{j}$ will also be perpendicular to $\hat{i} + \hat{j}$, so there are an infinite number of vectors in the plane that are perpendicular to $\hat{i} + \hat{j}$.

A simpler solution comes from noting that $\hat{k}$ is perpendicular to the x-y plane, so any scalar multiple of $\hat{k}$, $\alpha\hat{k}$ with $\alpha =$ any real number, is perpendicular to $\hat{i} + \hat{j}$.

2.45   First write the vector $\vec{\mathbf{B}}$ in Cartesian form:

$$\vec{\mathbf{B}} = 5.20 \cos 120° \hat{\mathbf{i}} + 5.20 \sin 120° \hat{\mathbf{j}} = 5.20(-0.500)\hat{\mathbf{i}} + 5.20(0.866)\hat{\mathbf{j}} = -2.60 \hat{\mathbf{i}} + 4.50 \hat{\mathbf{j}}.$$

a)

$$\vec{\mathbf{A}} \bullet \vec{\mathbf{B}} = (-3.20\hat{\mathbf{i}} + 2.40\hat{\mathbf{j}}) \bullet (-2.60\hat{\mathbf{i}} + 4.50\hat{\mathbf{j}}) = (-3.20)(-2.60) + (2.40)(4.50) = 19.1.$$

b)

$$\vec{\mathbf{A}} \times \vec{\mathbf{B}} = (-3.20\hat{\mathbf{i}} + 2.40\hat{\mathbf{j}}) \times (-2.60\hat{\mathbf{i}} + 4.50\hat{\mathbf{j}})$$
$$= \mathbf{0} + (-3.20)(4.50)(\hat{\mathbf{i}} \times \hat{\mathbf{j}}) + (2.40)(-2.60)(\hat{\mathbf{j}} \times \hat{\mathbf{i}}) + \mathbf{0}$$
$$= (-3.20)(4.50)\hat{\mathbf{k}} + (2.40)(-2.60)(-\hat{\mathbf{k}}) = -14.40\hat{\mathbf{k}} + 6.24\hat{\mathbf{k}} = -8.2\hat{\mathbf{k}}.$$

2.49

a)

$$\vec{\mathbf{A}} \bullet \vec{\mathbf{C}} = (\hat{\mathbf{i}} + 2\hat{\mathbf{j}} - \hat{\mathbf{k}}) \bullet (3\hat{\mathbf{j}} + 6\hat{\mathbf{k}}) = 0 + 6 - 6 = 0.$$

b)   We want to find a vector $\vec{\mathbf{B}}$ so that $\vec{\mathbf{A}} \times \vec{\mathbf{B}} = \vec{\mathbf{C}}$.

The result of a vector product is perpendicular to each of the vectors in the product. From part a) we see that $\vec{\mathbf{C}}$ is perpendicular to $\vec{\mathbf{A}}$. So that's good! To find $\vec{\mathbf{B}}$, we'll find another vector $\vec{\mathbf{B}}'$ perpendicular to $\vec{\mathbf{C}}$, and then multiply it by a scalar, if necessary, so that $\vec{\mathbf{A}} \times \vec{\mathbf{B}} = \vec{\mathbf{C}}$.

In order for $\vec{\mathbf{B}}'$ and $\vec{\mathbf{C}}$ to be perpendicular, their dot product should be zero, so

$$\vec{\mathbf{B}}' \bullet \vec{\mathbf{C}} = 0 \implies (B_x'\hat{\mathbf{i}} + B_y'\hat{\mathbf{j}} + B_z'\hat{\mathbf{k}}) \bullet (3\hat{\mathbf{j}} + 6\hat{\mathbf{k}}) = 0 \implies 0B_x' + 3B_y' + 6B_z' = 0.$$

Since the coefficient of B_x' is zero, B_x' can be anything. For simplicity, choose $B_x' = 0$. Also choose $B_y' = -2$ and $B_z' = 1$, so $\vec{\mathbf{B}}' = -2\hat{\mathbf{j}} + \hat{\mathbf{k}}$. Now compute $\vec{\mathbf{A}} \times \vec{\mathbf{B}}'$.

$$\vec{\mathbf{A}} \times \vec{\mathbf{B}}' = (\hat{\mathbf{i}} + 2\hat{\mathbf{j}} - \hat{\mathbf{k}}) \times (-2\hat{\mathbf{j}} + \hat{\mathbf{k}}) = -\hat{\mathbf{j}} - 2\hat{\mathbf{k}}.$$

Whoops! This is not $\vec{\mathbf{C}}$. But if we multiply it by -3, it is $\vec{\mathbf{C}}$. So set $\vec{\mathbf{B}} = (-3)\vec{\mathbf{B}}'$, and then

$$\vec{\mathbf{A}} \times \vec{\mathbf{B}} = \vec{\mathbf{A}} \times ((-3)\vec{\mathbf{B}}') = (-3)\vec{\mathbf{A}} \times \vec{\mathbf{B}}' = (-3)(-\hat{\mathbf{j}} - 2\hat{\mathbf{k}}) = 3\hat{\mathbf{j}} + 6\hat{\mathbf{k}} = \vec{\mathbf{C}}$$

as desired. Thus

$$\vec{\mathbf{B}} = (-3)\vec{\mathbf{B}}' = 6\hat{\mathbf{j}} - 3\hat{\mathbf{k}}$$

does the trick. The solution is not unique since we could have set B_x' equal to any number we wanted; that is, the x-component of $\vec{\mathbf{B}}$ need not be 0, but can be any number you choose.

An alternative, more straightforward, way to do this problem is to write out what the equation $\vec{\mathbf{A}} \times \vec{\mathbf{B}} = \vec{\mathbf{C}}$ means in terms of the components B_x, B_y, and B_z. When you do this, that is, when you equate each component of $\vec{\mathbf{A}} \times \vec{\mathbf{B}}$ with the corresponding component of $\vec{\mathbf{C}}$, you end up with three simultaneous linear equations in the three unknowns B_x, B_y, and B_z. These equations turn out to be linearly dependent but they are consistent. There are, therefore, an infinite number of solutions $\vec{\mathbf{B}}$. These solutions all lie on a straight line parallel to the x axis and running through the solution $6\hat{\mathbf{j}} - 3\hat{\mathbf{k}}$ that we found. This way of viewing the problem does require a knowledge of linear algebra.

c)   We showed in b) that the scalar product $\vec{\mathbf{B}}' \bullet \vec{\mathbf{C}}$ is zero, and since $\vec{\mathbf{B}} = (-3)\vec{\mathbf{B}}'$, $\vec{\mathbf{B}} \bullet \vec{\mathbf{C}} = 0$.

Alternatively, since the vector product $\vec{\mathbf{C}}$ is defined to be perpendicular to each of the vectors which form it, $\vec{\mathbf{B}} \bullet \vec{\mathbf{C}}$ must equal zero.

2.53

$$\vec{A} \times (\vec{B}_1 + \vec{B}_2) = (A_x\hat{i} + A_y\hat{j} + A_z\hat{k}) \times ((B_{1x} + B_{2x})\hat{i} + (B_{1y} + B_{2y})\hat{j} + (B_{1z} + B_{2z})\hat{k})$$

$$
\begin{aligned}
= \;& A_x(B_{1y} + B_{2y})\hat{k} - A_x(B_{1z} + B_{2z})\hat{j} \\
& - A_y(B_{1x} + B_{2x})\hat{k} + A_y(B_{1z} + B_{2z})\hat{i} \\
& + A_z(B_{1x} + B_{2x})\hat{j} - A_z(B_{1y} + B_{2y})\hat{i}
\end{aligned}
$$

$$
\begin{aligned}
= \;& (A_yB_{1z} - A_zB_{1y})\hat{i} + (A_yB_{2z} - A_zB_{2y})\hat{i} \\
& + (A_zB_{1x} - A_xB_{1z})\hat{j} + (A_zB_{2x} - A_xB_{2z})\hat{j} \\
& + (A_xB_{1y} - A_yB_{1x})\hat{k} + (A_xB_{2y} - A_yB_{2x})\hat{k}
\end{aligned}
$$

$$= \vec{A} \times \vec{B}_1 + \vec{A} \times \vec{B}_2.$$

2.57

a)

$$\vec{A}_1 + \vec{A}_2 - 2\vec{A}_3 = (4.0\hat{j}) + (1.0\hat{i} - 3.0\hat{j} + 2.0\hat{k}) - 2(2.0\hat{i} - 1.0\hat{j} + 3.0\hat{k}) = -3.0\hat{i} + 3.0\hat{j} - 4.0\hat{k}.$$

b) Let's save ourselves some work!

Since $\vec{A}_1 = 4.0\hat{j}$, the scalar product $\vec{A}_1 \bullet (\vec{A}_2 \times \vec{A}_3)$ is just 4.00 times the coefficient of $\hat{j}$ in $\vec{A}_2 \times \vec{A}_3$. That coefficient is $-(1.0)(3.0) + (2.0)(2.0) = 1.0$. Therefore

$$\vec{A}_1 \bullet (\vec{A}_2 \times \vec{A}_3) = (4.0)(1.0) = 4.0.$$

c) Let θ be the angle between $\vec{A}_2$ and $\vec{A}_3$. Then

$$\vec{A}_2 \bullet \vec{A}_3 = A_1A_2\cos\theta \implies \cos\theta = \frac{\vec{A}_2 \bullet \vec{A}_3}{A_2A_3}.$$

Calculate $\vec{A}_2 \bullet \vec{A}_3$, A_2, and A_3:

$$\vec{A}_2 \bullet \vec{A}_3 = (1.0)(2.0) + (-3.0)(-1.0) + (3.0)(3.0) = 11.0.$$
$$A_2 = \sqrt{(1.0)^2 + (-3)^2 + (2.0)^2} = 3.7.$$
$$A_3 = \sqrt{(2.0)^2 + (-1.0)^2 + (3.0)^2} = 3.7.$$

Therefore

$$\cos\theta = \frac{11.0}{(3.7)(3.7)} = 0.80 \implies \theta = 37°.$$

2.61

a) The magnitude of the vector product $\vec{B} \times \vec{C}$ is $|\vec{B} \times \vec{C}| = BC\sin\theta$, where θ is the angle between the vectors, as shown below:

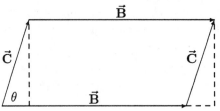

The area of a parallelogram is the product of it's base and height. For the above parallelogram, the base is B and the height is $C \sin \theta$. Therefore

$$\text{Area} = (B \sin \theta)\, C = |\vec{\mathbf{B}} \times \vec{\mathbf{C}}|.$$

b)   The direction of the *vector* $\vec{\mathbf{B}} \times \vec{\mathbf{C}}$ is perpendicular to the plane of $\vec{\mathbf{B}}$ and $\vec{\mathbf{C}}$ and goes along the vertical dashed line shown in figure P.61 of the text. Let ϕ be the angle between $\vec{\mathbf{A}}$ and the vertical dashed line. Then the height of the parallelepiped is $A \cos \phi$.

The volume of a parallelepiped is the product of the area of its base times its height. The area of the parallelogram that forms the base of the parallelepiped was shown in part a) to be $|\vec{\mathbf{B}} \times \vec{\mathbf{C}}|$. The height of the parallelepiped is $A \cos \phi$. Therefore the volume of the parallelepiped is

$$\text{Volume} = (|\vec{\mathbf{B}} \times \vec{\mathbf{C}}|)(A \cos \phi) = A|\vec{\mathbf{B}} \times \vec{\mathbf{C}}| \cos \phi.$$

But ϕ is the angle between the two vectors $\vec{\mathbf{A}}$ and $\vec{\mathbf{B}} \times \vec{\mathbf{C}}$. Hence

$$\vec{\mathbf{A}} \bullet (\vec{\mathbf{B}} \times \vec{\mathbf{C}}) = A|\vec{\mathbf{B}} \times \vec{\mathbf{C}}| \cos \phi = \text{Volume}.$$

2.65

$$\frac{d\vec{\mathbf{A}}}{dt} = \frac{d}{dt}(5\sqrt{t}\,\hat{\mathbf{i}} - 6t^{3/2}\,\hat{\mathbf{j}}) = 5\frac{d}{dt}\sqrt{t}\,\hat{\mathbf{i}} - 6\frac{d}{dt}t^{3/2}\hat{\mathbf{j}} = \frac{5}{2\sqrt{t}}\,\hat{\mathbf{i}} - 9\sqrt{t}\,\hat{\mathbf{j}}.$$

2.69

$$\frac{d}{dt}(\vec{\mathbf{A}} \times \vec{\mathbf{B}}) = \frac{d}{dt}((A_y B_z - A_z B_y)\hat{\mathbf{i}} + (A_z B_x - A_x B_z)\hat{\mathbf{j}} + (A_x B_y - A_y B_x)\hat{\mathbf{k}})$$

$$= \left(\frac{dA_y}{dt}B_z + A_y\frac{dB_z}{dt} - \frac{dA_z}{dt}B_y - A_z\frac{dB_y}{dt}\right)\hat{\mathbf{i}} +$$

$$\left(\frac{dA_z}{dt}B_x + A_z\frac{dB_x}{dt} - \frac{dA_x}{dt}B_z - A_x\frac{dB_z}{dt}\right)\hat{\mathbf{j}} +$$

$$\left(\frac{dA_x}{dt}B_y + A_x\frac{dB_y}{dt} - \frac{dA_y}{dt}B_x - A_y\frac{dB_x}{dt}\right)\hat{\mathbf{k}}$$

$$= \left(\frac{dA_y}{dt}B_z - \frac{dA_z}{dt}B_y\right)\hat{\mathbf{i}} + \left(A_y\frac{dB_z}{dt} - A_z\frac{dB_y}{dt}\right)\hat{\mathbf{i}} +$$

$$= \left(\frac{dA_z}{dt}B_x - \frac{dA_x}{dt}B_z\right)\hat{\mathbf{i}} + \left(A_z\frac{dB_x}{dt} - A_x\frac{dB_z}{dt}\right)\hat{\mathbf{j}} +$$

$$= \left(\frac{dA_x}{dt}B_y - \frac{dA_y}{dt}B_x\right)\hat{\mathbf{i}} + \left(A_x\frac{dB_y}{dt} - A_y\frac{dB_x}{dt}\right)\hat{\mathbf{k}}$$

$$= \frac{d\vec{\mathbf{A}}}{dt} \times \vec{\mathbf{B}} + \vec{\mathbf{A}} \times \frac{d\vec{\mathbf{B}}}{dt}.$$

Chapter 3

Kinematics I: Rectilinear Motion

3.1

a) Since the dog returns to the same point in space, the initial and final position vectors of the dog are the same. Hence $\Delta\vec{r} \equiv \vec{r}_{\text{final}} - \vec{r}_{\text{initial}} = \mathbf{0}$ m.

b) The total distance traveled by the dog is the round-trip distance: 25 m $+ 25$ m $= 50$ m.

c) The average speed of the dog is the total distance divided by the elapsed time: $v_{\text{average}} = 50$ m $/12$ s $= 4.2$ m/s.

d) The average velocity is the change $\Delta\vec{r}$ in the dog's position vector divided by the elapsed time: $\vec{v}_{\text{average}} = \dfrac{1}{12\text{ s}}\mathbf{0}$ m $= \mathbf{0}$ m/s.

3.5 The time t spent in the car is the total distance traveled divided by the average speed v_{average}:

$$t = \frac{d}{v_{\text{average}}} = \frac{4.0 \times 10^4 \text{ km}}{60 \text{ km/h}} = 6.7 \times 10^2 \text{ h}.$$

Converted to weeks, this is

$$6.7 \times 10^2 \text{ h} = (6.7 \times 10^2 \text{ h})\left(\frac{\text{d}}{24 \text{ h}}\right)\left(\frac{\text{week}}{7 \text{ d}}\right) = 4.0 \text{ week}.$$

3.9 The minimum speed is the distance divided by the reaction time: $\dfrac{2.0 \text{ m}}{0.30 \text{ s}} = 6.7$ m/s. Since the actual speed of the nerve impulses is on the order of 10^2 m/s, the time for them to propagate from your head to your foot is about $\dfrac{2.0 \text{ m}}{10^2 \text{ m/s}} = 2 \times 10^{-2}$ s.

3.13 First convert 30.0 km/h to meters per second and 15.00 min to seconds:

$$30.0 \text{ km/h} = \left(30.0\frac{\text{km}}{\text{h}}\right)\left(10^3\frac{\text{m}}{\text{km}}\right)\left(\frac{\text{h}}{3\,600 \text{ s}}\right) = 8.33 \text{ m/s}, \qquad 15.00 \text{ min} = 15.00 \text{ min}\left(60\frac{\text{s}}{\text{min}}\right) = 900.0 \text{ s}.$$

a) Choose $\hat{\mathbf{i}}$ to be in the direction of motion of the super tanker, and let $t = 0$ s be when the tanker starts to slow down. Then the initial velocity of the tanker is $\vec{v}_0 = (8.33 \text{ m/s})\hat{\mathbf{i}}$, so the equation of motion for the tanker is

$$\vec{v}(t) = (8.33 \text{ m/s})\hat{\mathbf{i}} + \vec{a}t.$$

When $t = 900$ s, the velocity is $\mathbf{0}$ m/s and the equation $v_x = v_{x0} + a_x t$ becomes

$$0 \text{ m/s} = (8.33 \text{ m/s}) + a_x(900 \text{ s}) \implies a_x = (-9.26 \times 10^{-3} \text{ m/s}^2).$$

15

b)   The equation for the x-component of the position, for motion with a constant acceleration a_x, is

$$x(t) = x_0 + v_{x0}t + a_x\frac{t^2}{2}.$$

Choose the origin of the coordinate system to be the point at which the tanker begins to slow down. Then

$$x_0 = 0 \text{ m}, \quad v_{x0} = (8.33 \text{ m/s}), \quad \text{and } a_x = (-9.26 \times 10^{-3} \text{ m/s}^2),$$

and

$$x(t) = 0 \text{ m/s} + (8.33 \text{ m/s})t + (-9.26 \times 10^{-3} \text{ m/s}^2)\frac{t^2}{2}.$$

Hence, when $t = 900$ s,

$$x(t) = 0 \text{ m/s} + (8.33 \text{ m/s})(900 \text{ s}) + (-9.26 \times 10^{-3} \text{ m/s}^2)\frac{(900 \text{ s})^2}{2} = (3.75 \times 10^3 \text{ m}).$$

So the tanker moved 3.75 km before stopping.

3.17

a)   Choose a coordinate system with origin on the Earth and $\hat{\mathbf{i}}$ pointed up. Then

$$x_0 = 30.0 \text{ m}, \quad v_{x0} = 0 \text{ m/s}, \quad \text{and} \quad a_x = -9.81 \text{ m/s}^2,$$

so

$$x(t) = x_0 + v_{x0}t + a_x\frac{t^2}{2} \implies 0 \text{ m} = 30 \text{ m} + (0 \text{ m/s})t - (9.81 \text{ m/s}^2)\frac{t^2}{2}.$$

Solve this to find $t = 2.47$ s.

b)   The maximum speed of the pot occurs at impact, when $t = 2.47$ s. The x-component of the velocity at this instant is

$$v_x(t) = 0 \text{ m/s} + (-9.81 \text{ m/s}^2)t = (-24.2 \text{ m/s}).$$

c)   When $x = 10.0$ m,

$$10.0 \text{ m} = 30.0 \text{ m} - (9.81 \text{ m/s}^2)\frac{t^2}{2} \implies t = 2.02 \text{ s}.$$

The x-component of the velocity at this time is

$$v_x(t) = 0 + (-9.81 \text{ m/s}^2)(2.02) = (-19.8 \text{ m/s}^2),$$

so the speed is $v = 19.8$ m/s.

d)   When $t = 1.00$ s, $x(1.00 \text{ s}) = 30.0 \text{ m} - (9.81 \text{ m/s}^2)\frac{(1.00 \text{ s})^2}{2} = 25.1 \text{ m}.$

3.21

a)   The velocity component at any instant is the slope of the position versus time graph at that instant. Use a straight edge to draw a tangent line to the curve at the point where $t = 3.0$ s. The slope of this line is $v_x = \dfrac{\Delta x}{\Delta t} \approx -6.0$ m/s.

b)   The velocity component is zero when the tangent line has zero slope, i.e., when it is horizontal. This occurs when $t \approx 1.5$ s.

c)   The particle is moving in the positive direction until $t \approx 1.5$ s, at which time its velocity becomes zero. After that, it moves in the negative direction. The *position* at which the reversal occurs is $x \approx 5.2$ m.

3.25   Let $\hat{\mathbf{i}}$ be directed along the direction in which the electrons are moving. The average acceleration of an electron is

$$\vec{\mathbf{a}}_{\text{ave}} = \frac{\Delta \vec{\mathbf{v}}}{\Delta t} = \left(\frac{0 \text{ m/s} - 3 \times 10^6 \text{ m/s}}{1 \times 10^{-18} \text{ s}} \right) \hat{\mathbf{i}} = (-3 \times 10^{24} \text{ m/s}^2) \hat{\mathbf{i}}.$$

The magnitude of this vector is 3×10^{24} m/s^2 ! That's acceleration!

3.29   Set up a coordinate system whose origin is at the place where the brakes are first applied and with $\hat{\mathbf{i}}$ in the direction of motion of the car. We are told that $x_0 = 0$ m, $v_{x0} = 30.0$ m/s, and that when $t = 5.0$ s, $v_x = 15.0$ m/s. Substitute these values into the general equation $v_x(t) = v_{x0} + a_x t$ and solve for a_x to find $a_x = -3.0$ m/s^2.

a)   The time the car stops can now be found from the velocity component equation:

$$v_x(t) = 30.0 \text{ m/s} + (-3.0 \text{ m/s}^2)t.$$

The car is stopped when the left-hand side is zero, so

$$0 \text{ m/s} = 30.0 \text{ m/s} + (-3.0 \text{ m/s}^2)t \implies t = 10 \text{ s}.$$

b)   The position of the car when $t = 10$ s is found from

$$x(10 \text{ s}) = 0 \text{ m} + (30.0 \text{ m/s})(10 \text{ s}) - (3.0 \text{ m/s}^2)\frac{(10 \text{ s})^2}{2} = 150 \text{ m}.$$

Since the car began braking 160 m from the bridge, it is a comfortable 10 m from the bridge when it stops.

3.33

a)   The average acceleration is the change in the velocity divided by the time interval during which the change took place:

$$\vec{\mathbf{a}}_{\text{ave}} = \frac{\Delta \vec{\mathbf{v}}}{\Delta t} = \frac{(2.0 \text{ m/s})\hat{\mathbf{i}} - (0 \text{ m/s})\hat{\mathbf{i}}}{60 \text{ s}} = (3.3 \times 10^{-2} \text{ m/s}^2)\hat{\mathbf{i}}.$$

The instantaneous acceleration when $t = 0.5$ min is the slope of the v_x versus t graph at that instant. Since the graph is a straight line between 0 min and 1 min, the instantaneous acceleration at each instant between these times is the same as the average acceleration during this interval. Hence, the instantaneous acceleration when $t = 0.5$ min is $(3.3 \times 10^{-2} \text{ m/s}^2)\hat{\mathbf{i}}$.

b)   The average acceleration is the change in the velocity divided by the time interval during which the change took place:

$$\vec{\mathbf{a}}_{\text{ave}} = \frac{\Delta \vec{\mathbf{v}}}{\Delta t} = \frac{(-2.0 \text{ m/s})\hat{\mathbf{i}} - (2.0 \text{ m/s})\hat{\mathbf{i}}}{60 \text{ s}} = (-6.7 \times 10^{-2} \text{ m/s}^2)\hat{\mathbf{i}}.$$

The instantaneous acceleration when $t = 1.5$ min is the slope of the v_x versus t graph at that instant. Since the graph is a straight line between 1 min and 2 min, the instantaneous acceleration at each instant between these times is the same as the average acceleration during this interval. Hence, the instantaneous acceleration when $t = 0.5$ min is $(-6.7 \times 10^{-2} \text{ m/s}^2)\hat{\mathbf{i}}$.

c)   The average acceleration is the change in the velocity divided by the time interval during which the change took place:

$$\vec{\mathbf{a}}_{\text{ave}} = \frac{\Delta \vec{\mathbf{v}}}{\Delta t} = \frac{(0 \text{ m/s})\hat{\mathbf{i}} - (-2.0 \text{ m/s})\hat{\mathbf{i}}}{60 \text{ s}} = (3.3 \times 10^{-2} \text{ m/s}^2)\hat{\mathbf{i}}.$$

The instantaneous acceleration when $t = 2.5$ min is the slope of the v_x versus t graph at that instant. Since the graph is a straight line between 2 min and 3 min, the instantaneous acceleration at each instant between these times is the same as the average acceleration during this interval. Hence, the instantaneous acceleration when $t = 2.5$ min is $(3.3 \times 10^{-2} \text{ m/s}^2)\hat{\mathbf{i}}$.

d)   The barge begins at rest when $t = 0$ s and accelerates to 2.0 m/s when $t = 60$ s. The barge then begins to slow down until it is instantaneously at rest when $t = 90$ s. It then reverses direction until it reaches a speed of 2.0 m/s when $t = 120$ s. The barge then slows down and stops when $t = 180$ s.

e)   During the first 60 seconds the acceleration component is $a_x = 3.3 \times 10^{-2}$ m/s^2. When $t = 0$ s, $x_0 = 0$ m, $v_{x0} = 0$ m/s, hence

$$x(t) = x_0 + v_{x0}t + a_x\frac{t^2}{2} = (3.3 \times 10^{-2} \text{ m/s}^2)\frac{t^2}{2}.$$

This is the equation of a parabola with vertex at the origin, opening upwards. (A happy parabola!) Here are the values of x corresponding to various values of t:

t (s)	x (m)
0	0
10	1.7
20	6.6
30	15
40	26
50	41
60	59

The next time interval of constant acceleration begins at $t_1 = 60$ s and ends at $t_2 = 120$ s. To find $x(t)$ in this interval, we replace $(t - 0)$ by $(t - t_1)$ and v_{x0} by $v_x(t_1)$ in the general equation for position, giving

$$x(t) = x(t_1) + v_x(t_1)(t - t_1) + a_x\frac{(t - t_1)^2}{2},$$

where now $a_x = -6.7 \times 10^{-2}$ m/s^2. The initial position is $x(60) = 59$ m, and the initial velocity component is the final velocity component from the previous time interval:

$$v_x(60 \text{ s}) = 0 \text{ m/s} + (3.3 \times 10^{-2} \text{ m/s}^2)(60 \text{ s}) = 2.0 \text{ m/s}.$$

So, substituting these values into the new equation for $x(t)$:

$$x(t) = 59 \text{ m} + (2.0 \text{ m/s})(t - 60 \text{ s}) + (-6.7 \times 10^{-2} \text{ m/s}^2)\frac{(t - 60 \text{ s})^2}{2}.$$

From this equation we find:

t (s)	x (m)
60	59
70	76
80	86
90	89
100	86
110	76
120	59

During the third minute, the acceleration component returns to what it was during the first minute, and the barge has positions which are the reverse of those during the first minute. Hence the graph of the position of the barge as a function of time during the three minute voyage is:

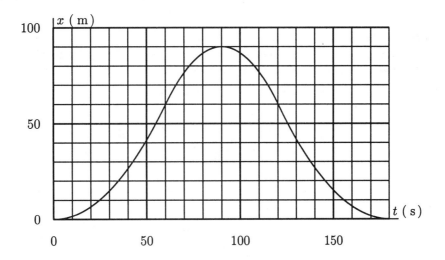

3.37

a)   The animal is at rest when its velocity component is zero. From the graph this is when $t = 0$ s and again when $t = 6.0$ s.

b)   The acceleration component is the slope of the v_x versus t graph. The acceleration component is zero when the slope is zero, i.e., during the interval from $t = 1.0$ s to $t = 4.0$ s.

c)   The acceleration component when $t = 4.5$ s is the slope of the v_x versus t graph at that instant. This slope is approximately

$$a_x = \frac{\Delta v_x}{\Delta t} = \frac{0 \text{ m/s} - 4.0 \text{ m/s}}{6.0 \text{ s} - 4.0 \text{ s}} = -2.0 \text{ m/s}^2 \,.$$

Hence the acceleration is $\vec{a} = (-2.0 \text{ m/s}^2)\hat{\mathbf{i}}$.

d)   The change in the position coordinate from $t = 0$ s to 6.0 s is the total area under the v_x versus t graph between 0 s and 6.0 s:

	triangular area		rectangular area		triangular area
$x(6\text{ s}) =$	between	$+$	between	$+$	between
	0 s and 1.0 s		1.0 s and 4.0 s		4.0 s and 6.0 s

$$= \frac{1}{2}(1.0 \text{ s})(4.0 \text{ m/s}) \;+\; (3.0 \text{ s})(4.0 \text{ m/s}) \;+\; \frac{1}{2}(2.0 \text{ s})(4.0 \text{ m/s})$$

$$= 18 \text{ m} \,.$$

This result also can be obtained as follows:

During the first second, the acceleration component of the kudu is constant and equal to $a_x = 4.0 \text{ m/s}^2$. Its position and velocity components at time $t = 0$ s are $x(0\text{ s}) = 0$ m and $v_x(0\text{ s}) = 0 \text{ m/s}^2$. Hence, during the first second of travel its position is given by $x(t) = 0 \text{ m} + (0 \text{ m/s})t + (4.0 \text{ m/s}^2)\frac{t^2}{2} = (4.0 \text{ m/s}^2)\frac{t^2}{2}$. When $t = 0$ s this gives the kudu's position as $(4.0 \text{ m/s}^2)\frac{(1 \text{ s})^2}{2} = 2 \text{ m}$.

For the next 3.0 s, the kudu travels with a constant velocity component equal to 4.0 m/s, so it covers an additional $(4.0 \text{ m/s})(3.0 \text{ s}) = 12 \text{ m}$. Thus it is located 14 m from the origin when $t = 4.0$ s.

During the final 2.0 s, the kudu begins the interval with a velocity component of 4.0 m/s and has an acceleration component of -2.0 m/s^2, so it travels an additional $(4.0 \text{ m/s})(2.0 \text{ s}) + (-2.0 \text{ m/s}^2)\frac{(2.0 \text{ s})^2}{2} = 4.0 \text{ m}$, bringing it a total of 18 m from the origin.

3.41

a)   Choose $\hat{\mathbf{i}}$ to be along the direction of the arrow with the origin at the rear of the arrow, and let $t = 0$ at the time of release. Then $v_{x0} = 0$ m/s and $x_0 = 0$ m, so for constant acceleration, the equations for $v_x(t)$ and $x(t)$ are

$$v_x(t) = a_x t, \qquad \text{and} \qquad x(t) = a_x \frac{t^2}{2}.$$

At the time t_f, when the arrow leaves the bow, these equations become

$$30 \text{ m/s} = a_x t_f, \qquad \text{and} \qquad 0.70 \text{ m} = a_x \frac{t_f^2}{2}.$$

These equations may be solved simultaneously for a_x and t_f: Solve the first one for t_f, $t_f = \dfrac{30 \text{ m/s}}{a_x}$, then substitute this value for t_f into the second:

$$0.70 \text{ m} = a_x \frac{\left(\dfrac{30 \text{ m/s}}{a_x}\right)^2}{2} = 4.5 \times 10^2 \left(\frac{\text{m} \cdot \text{m}}{\text{s} \cdot \text{s}}\right) \frac{1}{a_x}.$$

Therefore, $a_x = 6.4 \times 10^2$ m/s^2.

b)   Change coordinate systems. Let $\hat{\mathbf{i}}$ point straight up and let the origin be at ground level. Let $t = 0$ s be the time that the arrow leaves the bow. Then $x_0 = 2$ m, $v_{x0} = 30$ m/s, and $a_x = -9.81$ m/s^2, so the equations for $v_x(t)$ and $x(t)$ are

$$v_x(t) = 30 \text{ m/s} - (9.81 \text{ m/s}^2)t, \qquad \text{and} \qquad x(t) = 2.0 \text{ m} + (30 \text{ m/s})t - (9.81 \text{ m/s}^2)\frac{t^2}{2}.$$

At the time t_f that the arrow is at maximum height, $v_x(t_f) = 0$ m/s. So from the first equation, 0 m/s = 30 m/s $-$ (9.81 m/s^2)t_f. Solve this to get $t_f = 3.1$ s. Substitute this value for t_f into the equation for $x(t)$ to get $x(t_f) = 2.0 \text{ m} + (30 \text{ m/s})(3.1 \text{ s}) - (9.81 \text{ m/s}^2)\dfrac{(3.1 \text{ s})^2}{2} = 48$ m. So the arrow reaches a maximum height of 48 m.

Now let t_f denote the total flight time (the time when the arrow hits the ground). Then $x(t_f) = 0$ m, so the equation for $x(t)$ becomes

$$0 \text{ m} = 2.0 \text{ m} + (30 \text{ m/s})t_f - (9.81 \text{ m/s}^2)\frac{t_f^2}{2}.$$

Use the quadratic equation to solve this for the positive root, and find that $t_f = 6.2$ s.

3.45

a)   Choose $\hat{\mathbf{i}}$ to be in the direction of the motion of the train with the origin at the initial position of the rear of the train, so initially the commuters' position vector is $(-30.0 \text{ m})\hat{\mathbf{i}}$, and the position vector for the rear of the train is $\mathbf{0}$ m.

b)   Since the train's initial position and velocity are both zero, the train's position coordinate at time t is

$$x_{\text{train}}(t) = (0.250 \text{ m/s}^2)\frac{t^2}{2}.$$

The commuters' initial position coordinate is -30.0 m, their initial velocity is $(4.00 \text{ m/s})\hat{\mathbf{i}}$, and their acceleration is zero. Hence, at any time t, their position coordinate is

$$x_{\text{commuters}}(t) = -30.0 \text{ m} + (4.00 \text{ m/s})t.$$

To find the time(s) when they can jump aboard we set $x_{\text{train}}(t) = x_{\text{commuters}}(t)$, which gives us the quadratic equation

$$(0.250 \text{ m/s}^2)\frac{t^2}{2} = -30.0 \text{ m} + (4.00 \text{ m/s})t.$$

Using the quadratic formula, we find this equation has two solutions,

$$t = 12.0 \text{ s} \quad \text{and} \quad t = 20.0 \text{ s}.$$

c) The first solution, $t = 12.0$ s, is the instant they catch up to the train. If they don't jump aboard, first they will pass the rear of the train, and then the rear of the train catches up to them when $t = 20.0$ s. This is their last chance to jump aboard.

3.49 Choose a coordinate system with $\hat{\mathbf{i}}$ in the direction of travel of the electrons and the origin at the point where each electron begins its journey along the tube. Then $x_0 = 0$ m and $v_{x0} = 0$ m/s. So, the equations for the velocity component and position are

$$v_x = a_x t \quad \text{and} \quad x(t) = a_x \frac{t^2}{2}.$$

When the electron reaches the end of the tube at time t_f, its velocity component is $v_x(t_f) = 6.0 \times 10^6$ m/s, and its position is $x(t_f) = 3.0 \times 10^3$ m, so these equations become

$$6.0 \times 10^6 \text{ m/s} = a_x t_f \quad \text{and} \quad 3.0 \times 10^3 \text{ m} = a_x \frac{t_f^2}{2}.$$

Solve the first equation for a_x in terms of t_f, $a_x = \dfrac{6.0 \times 10^6 \text{ m/s}}{t_f}$, and then substitute this expression for a_x into the second equation: $3.0 \times 10^3 \text{ m} = \left(\dfrac{6.0 \times 10^6 \text{ m/s}}{t_f}\right)\left(\dfrac{t_f^2}{2}\right)$. Solve this equation for t_f to get $t_f = 1.0 \times 10^{-3}$ s. Then use this to find a_x: $a_x = \dfrac{6.0 \times 10^6 \text{ m/s}}{t_f} = 6.0 \times 10^9 \text{ m/s}^2$.

3.53

a) Choose a coordinate system with $\hat{\mathbf{i}}$ pointing up and with origin at the point where the ball is released.
b) With this coordinate system, we have $x_0 = 0$ m, $v_{x0} = 25.0$ m/s, and $a_x = -9.81$ m/s^2. The equation for the x-component of the velocity is

$$v_x(t) = v_{x0} + a_x t = 25.0 \text{ m/s} - (9.81 \text{ m/s}^2)t.$$

When the ball reaches its maximum height, its velocity is zero; hence

$$0 \text{ m/s} = 25.0 \text{ m/s} - (9.81 \text{ m/s}^2)t.$$

Solve this for t:

$$t = 2.55 \text{ s}.$$

The equation for the position of the ball is

$$x(t) = x_0 + v_{x0}t + a_x\frac{t^2}{2} = 0 \text{ m} + (25.0 \text{ m/s})t - (9.81 \text{ m/s}^2)\frac{t^2}{2}.$$

To find its maximum altitude, let $t = 2.55$ s:

$$x(2.55 \text{ s}) = (25.0 \text{ m/s})(2.55 \text{ s}) - (9.81 \text{ m/s}^2)\frac{(2.55 \text{ s})^2}{2} = 31.9 \text{ m}.$$

c)   Since the ball returns to its initial position, the time the ball rises is equal to the time it falls, so the total time is $2(2.55 \text{ s}) = 5.10 \text{ s}$. (This also can be found by setting $x = 0$ m in the equation for the position and solving for t.)

d)   The acceleration component is constant at -9.81 m/s^2 throughout the flight. Its graph is

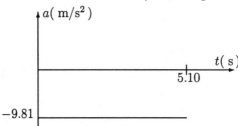

e)   The velocity component is given by the equation $v_x = 25.0 \text{ m/s} + (-9.81 \text{ m/s})t$. Its graph is

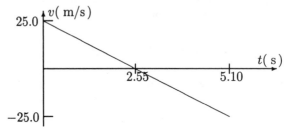

f)   The x-component of the position is $x(t) = (25.0 \text{ m/s})t - (9.81 \text{ m/s}^2)\dfrac{t^2}{2}$. Its graph is

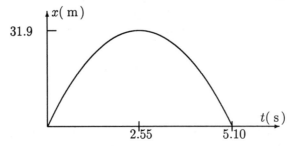

It has a maximum at the point $(2.55 \text{ s}, 31.9 \text{ m})$.

3.57   Choose a coordinate system with $\hat{\mathbf{i}}$ pointing up and with origin on the ground. Then $v_{x0} = 3.0$ m/s and $a_x = -9.81$ m/s^2. The equation for the x-component of position is

$$x(t) = x_0 + v_{x0}t + a_x\frac{t^2}{2} = x_0 + (3.0 \text{ m/s})t + (9.81 \text{ m/s}^2)\frac{t^2}{2}.$$

When $t = 1.5$ s, the pine cone is at the origin, so

$$0 \text{ m} = x_0 + (3.0 \text{ m/s})(1.5 \text{ s}) + (-9.81 \text{ m/s}^2)\frac{(1.5 \text{ s})^2}{2}.$$

Solving for x_0, $x_0 = 6.5$ m.

3.61   Choose a coordinate system with $\hat{\mathbf{i}}$ pointing up and the origin at the point of impact. With this choice $x_0 = 0$ m, $v_{x0} = -50$ m/s, and $a_x = 75(9.81 \text{ m/s}^2) = 736$ m/s^2. The equation for the velocity component is

$$v_x(t) = -50 \text{ m/s} + (736 \text{ m/s}^2)t.$$

9

When the sky diver is brought to rest, $v_x(t) = 0$. Solve the above equation for v_x for the time necessary for this to occur: $t = \dfrac{50 \text{ m/s}}{736 \text{ m/s}^2} = 0.068 \text{ s}$.

The equation for position is

$$x(t) = (-50 \text{ m/s})t + (736 \text{ m/s}^2)\frac{t^2}{2}.$$

When t = 0.068 (s), the sky diver is at rest. Substitute this value into the above equation for $x(t)$ to find $x = -1.7$ m. Thus, the sky diver is brought to rest over a distance of 1.7 m.

3.65   Take $\hat{\mathbf{i}}$ to point down and let the origin be at the top of the well. Let the (unknown) depth of the well be d. For a rock falling from rest we have $x_0 = 0$ m, $v_{x0} = 0$ m/s, and $a_x = g$. Using these initial values, the x-component of the position vector of the ball at the time t when it reaches the bottom of the well is $d = g\dfrac{t^2}{2}$, so

$$t = \sqrt{\frac{2d}{g}}.$$

The time t' for the sound of the splash to propagate from the bottom to the top of the well is $t' = \dfrac{d}{v_s}$ where v_s is the speed of sound. Thus, the total time T, from the release of the rock until the splash is heard at the top of the well, is

$$T = t + t' = \sqrt{\frac{2d}{g}} + \frac{d}{v_s}.$$

The time T is given as 3.00 s, hence

$$3.00 \text{ s} = \sqrt{\frac{2d}{9.81 \text{ m/s}^2}} + \frac{d}{343 \text{ m/s}}$$

Let $\alpha = \sqrt{d}$. Then in terms of α

$$3.00 \text{ s} = \left(0.451\frac{\text{s}}{\text{m}^{1/2}}\right)\alpha + \frac{\alpha^2}{343 \text{ m/s}}.$$

Clear fractions and transpose:

$$\alpha^2 + (155 \text{ m}^{1/2})\alpha - 1.03 \times 10^3 \text{ m} = 0 \text{ m}.$$

This is quadratic in α. Use the quadratic formula to solve for the positive root:

$$\alpha = 6.5 \text{ m}^{1/2}.$$

Since $\alpha^2 = d$, the depth of the well is

$$d = (6.5 \text{ m}^{1/2})^2 = 42 \text{ m}.$$

3.69

a)   The acceleration component is the slope of the v_x versus t graph. This slope is negative. Hence, $\hat{\mathbf{i}}$ was chosen to point up.

b)   Since the velocity component initially is positive, the particle was moving upward when $t = 0$ s.

c)   The particle was at its greatest elevation when the velocity component was zero, i.e., when $t \approx 3.0$ s.

d)   The area under the curve between 0 s and 5.0 s is the sum of the two triangular areas, the first between 0 s and 3.0 s, and the second between 3.0 s and 5.0 s:

$$\frac{(3.0 \text{ s})(30.0 \text{ m/s})}{2} + \frac{(2.0 \text{ s})(-20.0 \text{ m/s})}{2} = 25 \text{ m}.$$

e)   The area under the v_x versus t curve between 0 s and 5.0 s is the change in x-component of the position vector during that interval, which was calculated in part (d) to be 25 m.

Chapter 4

Kinematics II: Motion in Two and Three Dimensions

4.1

a) Choose $\hat{\mathbf{i}}$ pointing to the right, $\hat{\mathbf{j}}$ pointing up, and the origin at the crook of the branch. Then the initial position vector is

$$\vec{\mathbf{r}}_i = (-0.15 \text{ m})\hat{\mathbf{i}}.$$

The final position vector has both horizontal and vertical components:

$$\vec{\mathbf{r}}_f = (0.30 \text{ m}) \cos 30°\hat{\mathbf{i}} + (0.30 \text{ m}) \sin 30°\hat{\mathbf{j}} = 0.26 \text{ m}\,\hat{\mathbf{i}} + 0.15 \text{ m}\,\hat{\mathbf{j}}.$$

b) The average speed is the total distance traveled divided by the time interval to traverse that distance:

$$\text{average speed} = \frac{\text{total distance}}{\Delta t} = \frac{0.15 \text{ m} + 0.30 \text{ m}}{60 \text{ s}} = 7.5 \times 10^{-3} \text{ m/s}.$$

The average velocity is the change in the position vector divided by the time interval to accomplish that change:

$$\vec{\mathbf{v}}_{\text{ave}} = \frac{\Delta \vec{\mathbf{r}}}{\Delta t} = \frac{(0.26 \text{ m})\hat{\mathbf{i}} + (0.15 \text{ m})\hat{\mathbf{j}} - (-0.15 \text{ m})\hat{\mathbf{i}}}{60 \text{ s}} = (6.8 \times 10^{-3} \text{ m/s})\hat{\mathbf{i}} + (2.5 \times 10^{-3} \text{ m/s})\hat{\mathbf{j}}.$$

The magnitude of the average velocity is:

$$v_{\text{ave}} = \sqrt{(6.8 \times 10^{-3} \text{ m/s})^2 + (2.5 \times 10^{-3} \text{ m/s})^2} = 7.2 \times 10^{-3} \text{ m/s}.$$

4.5

a) Note that the position vectors are related by $\vec{\mathbf{r}}_1 + \vec{\mathbf{r}} = \vec{\mathbf{r}}_2$, so $\vec{\mathbf{r}} = \vec{\mathbf{r}}_2 - \vec{\mathbf{r}}_1$. Thus,

$$r^2 = \vec{\mathbf{r}} \bullet \vec{\mathbf{r}} = (\vec{\mathbf{r}}_2 - \vec{\mathbf{r}}_1) \bullet (\vec{\mathbf{r}}_2 - \vec{\mathbf{r}}_1) = \vec{\mathbf{r}}_2 \bullet \vec{\mathbf{r}}_2 + \vec{\mathbf{r}}_1 \bullet \vec{\mathbf{r}}_1 - 2\vec{\mathbf{r}}_1 \bullet \vec{\mathbf{r}}_2$$
$$= r_2^2 + r_1^2 - 2r_1 r_2 \cos\theta = (v_2 t)^2 + (v_1 t)^2 - 2(v_1 t)(v_2 t) \cos\theta$$
$$= (v_1^2 + v_2^2 - 2v_1 v_2 \cos\theta)t^2.$$

Therefore $r = \sqrt{v_1^2 + v_2^2 - 2v_1 v_2 \cos\theta}\ t$.

b) Using the last result,

$$v = \frac{dr}{dt} = \frac{d}{dt}\left(\sqrt{v_1^2 + v_2^2 - 2v_1 v_2 \cos\theta}\ t\right) = \sqrt{v_1^2 + v_2^2 - 2v_1 v_2 \cos\theta}.$$

Note that with the expression for r above, the speed can be written as $v = \dfrac{r}{t}$.

4.9

a)   Choose a coordinate system with $\hat{\mathbf{i}}$ pointing right, $\hat{\mathbf{j}}$ pointing up, and origin at the launch point of the soccer ball.

b)   In this coordinate system,

$$
\begin{aligned}
y_0 &= 0 \text{ m} \\
v_{y0} &= (20.0 \text{ m/s}) \sin 30.0^\circ \\
&= 10.0 \text{ m/s} \\
a_y &= -g
\end{aligned}
\qquad\qquad
\begin{aligned}
x_0 &= 0 \text{ m} \\
v_{x0} &= (20.0 \text{ m/s}) \cos 30.0^\circ \\
&= 17.3 \text{ m/s} \\
a_x &= 0 \text{ m/s}^2 .
\end{aligned}
$$

The equations for the velocity and position components are

$$
\begin{aligned}
v_y(t) &= 10.0 \text{ m/s} - gt \\
y(t) &= (10.0 \text{ m/s})t - g\frac{t^2}{2}
\end{aligned}
\qquad\qquad
\begin{aligned}
v_x(t) &= 17.3 \text{ m/s} \\
x(t) &= (17.3 \text{ m/s})t.
\end{aligned}
$$

c)   The soccer ball impacts when $y(t) = -40.0$ m. Substitute this into the equation for $y(t)$,

$$
-40.0 \text{ m} = (10.0 \text{ m/s})t - g\frac{t^2}{2},
$$

and then use the quadratic formula to solve for t. The two roots are $t = 4.05$ s and $t = -2.01$ s. Since impact occurs after $t = 0$ (the time when the ball is kicked), choose the positive root. Thus, the flight time is 4.05 s.

d)   The x-coordinate of the impact point is determined from the equation for $x(t)$ with $t = 4.05$ s.

$$
x(t) = (17.3 \text{ m/s})t = (17.3 \text{ m/s})(4.05) = 70.1 \text{ m}.
$$

The coordinates of the impact point are therefore $x = 70.1$ m and $y = -40.0$ m.

4.13

a)   Choose a coordinate system with $\hat{\mathbf{i}}$ pointing to the right, $\hat{\mathbf{j}}$ pointing up, and origin at the launch point. In this coordinate system

$$
\begin{aligned}
y_0 &= 0 \text{ m} \\
v_{y0} &= (15.00 \text{ m/s}) \sin 60^\circ \\
&= 13.0 \text{ m/s} \\
a_y &= -g
\end{aligned}
\qquad\qquad
\begin{aligned}
x_0 &= 0 \text{ m} \\
v_{x0} &= (15.0 \text{ m/s}) \cos 60^\circ \\
&= 7.50 \text{ m/s} \\
a_x &= 0 \text{ m/s}^2 .
\end{aligned}
$$

So, the equations for the velocity components are

$$
v_y(t) = 13.0 \text{ m/s} - gt
\qquad\qquad
v_x(t) = 7.50 \text{ m/s}.
$$

When the ball is rising and the velocity vector makes an angle of 45° to the horizontal direction, the components of the velocity are equal to each other. So at this time

$$
13.0 \text{ m/s} - gt = 7.50 \text{ m/s}.
$$

Solve for t.

$$
t = 0.56 \text{ s}.
$$

When the ball is falling and the velocity vector makes an angle of 45° to the horizontal direction, the components of the velocity are the negatives of each other. So at this time

$$
13.0 \text{ m/s} - gt = -7.50 \text{ m/s}.
$$

Solve for t.

$$t = 2.09 \text{ s}.$$

4.17

a)   Choose a coordinate system with $\hat{\mathbf{i}}$ pointing to the right, $\hat{\mathbf{j}}$ pointing up, and origin at the launch point.

b)   In this coordinate system

$$
\begin{aligned}
y_0 &= 0 \text{ m} \\
v_{y0} &= v_0 \sin 30° \\
a_y &= -g
\end{aligned}
\qquad\qquad
\begin{aligned}
x_0 &= 0 \text{ m} \\
v_{x0} &= v_0 \cos 30° \\
a_x &= 0 \text{ m/s}^2.
\end{aligned}
$$

So, the equations for the velocity and position components are

$$
\begin{aligned}
v_y(t) &= v_0 \sin 30° - gt \\
y(t) &= v_0(\sin 30°)t - g\frac{t^2}{2}
\end{aligned}
\qquad\qquad
\begin{aligned}
v_x(t) &= v_0 \cos 30° \\
x(t) &= v_0(\cos 30°)t
\end{aligned}
$$

The x-coordinate of the impact is 8.00 m. Thus, at the time of impact,

$$8.00 \text{ m} = v_0(\cos 30.0°)t,$$

so the impact time is

(1)
$$t = \frac{8.00 \text{ m}}{v_0 \cos 30°} = \frac{9.24 \text{ m}}{v_0}.$$

The y-coordinate of the impact is 1.00 m. Hence, at the time of impact,

$$1.00 \text{ m} = v_0(\sin 30.0°)t - g\frac{t^2}{2}.$$

Substitute the impact time from equation (1) into this equation and find

$$1.00 \text{ m} = v_0(\sin 30.0°)\frac{9.24 \text{ m}}{v_0} - g\frac{\left(\dfrac{9.24 \text{ m}}{v_0}\right)^2}{2} = 4.62 \text{ m} - g\frac{\left(\dfrac{9.24 \text{ m}}{v_0}\right)^2}{2}.$$

Solve for v_0 to find $v_0 = 10.8 \text{ m/s}$.

c)   Use equation (1) and v_0 from part b) to find the time for the divot to reach the golf cart:

$$t = \frac{9.24 \text{ s}}{v_0} = 0.856 \text{ s}.$$

4.21

a)   Choose a coordinate system with $\hat{\mathbf{i}}$ pointing to the right, $\hat{\mathbf{j}}$ pointing up, and origin on the ground directly below the launch point.

b)   In this coordinate system

$$
\begin{aligned}
y_0 &= 15 \text{ m} \\
v_{y0} &= v_0 \sin 45° = \frac{v_0}{\sqrt{2}} \\
a_y &= -g
\end{aligned}
\qquad\qquad
\begin{aligned}
x_0 &= 0 \text{ m} \\
v_{x0} &= v_0 \cos 45° = \frac{v_0}{\sqrt{2}} \\
a_x &= 0 \text{ m/s}^2.
\end{aligned}
$$

So, the equations for the velocity and position components are

$$v_y(t) = \frac{v_0}{\sqrt{2}} - gt \qquad\qquad\qquad v_x(t) = \frac{v_0}{\sqrt{2}}$$

$$y(t) = 15 \text{ m} + \frac{v_0}{\sqrt{2}}t - g\frac{t^2}{2} \qquad\qquad x(t) = \frac{v_0}{\sqrt{2}}t.$$

The x-coordinate of the upright piano's impact point is 140 m. Use this information in the equation for $x(t)$ at the time of impact,

$$140 \text{ m} = \frac{v_0}{\sqrt{2}}t$$

and solve for the impact time:

$$t = \frac{(140 \text{ m})\sqrt{2}}{v_0}.$$

The y-coordinate of the impact point is 0 m. Hence, the equation for $y(t)$ yields

$$0 \text{ m} = 15 \text{ m} + \frac{v_0}{\sqrt{2}}t - g\frac{t^2}{2}.$$

Substitute the expression for the impact time into this equation,

$$0 \text{ m} = 15 \text{ m} + \frac{v_0}{\sqrt{2}}\left(\frac{(140 \text{ m})\sqrt{2}}{v_0}\right) - g\frac{\left(\frac{(140 \text{ m})\sqrt{2}}{v_0}\right)^2}{2} = 155 \text{ m} - g\frac{\left(\frac{(140 \text{ m})\sqrt{2}}{v_0}\right)^2}{2},$$

and solve for v_0:

$$v_0 = 35 \text{ m/s}.$$

The average acceleration is the change in velocity divided by the time required to make that change:

$$\vec{\mathbf{v}}_{\text{ave}} = \frac{\Delta\vec{\mathbf{v}}}{\Delta t} = \frac{\vec{\mathbf{v}}_\text{f} - \vec{\mathbf{v}}_\text{i}}{\Delta t} = \frac{(35 \text{ m/s})(\cos 45°)\hat{\mathbf{i}} + (35 \text{ m/s})(\sin 45°)\hat{\mathbf{j}} - \mathbf{0} \text{ m/s}}{1.5 \text{ s}} = (16 \text{ m/s}^2)\hat{\mathbf{i}} + (16 \text{ m/s}^2)\hat{\mathbf{j}}.$$

The magnitude of the average acceleration is

$$v_{\text{ave}} = \sqrt{(16 \text{ m/s}^2)^2 + (16 \text{ m/s}^2)^2} = 23 \text{ m/s}^2.$$

4.25 Using the coordinate system given in the problem

$$y_0 = 0 \text{ m} \qquad\qquad\qquad x_0 = 0 \text{ m}$$
$$v_{y0} = v_0 \sin\theta \qquad\qquad\qquad v_{x0} = v_0 \cos\theta$$
$$a_y = -g \qquad\qquad\qquad a_x = 0 \text{ m/s}^2.$$

So, the kinematics equations are

$$v_y(t) = v_0 \sin\theta - gt \qquad\qquad\qquad v_x(t) = v_0 \cos\theta$$
$$y(t) = (v_0 \sin\theta)t - g\frac{t^2}{2} \qquad\qquad\quad x(t) = (v_0 \cos\theta)t.$$

Let R be the range of the projectile up the slope. Then the x-coordinate of the impact point is $R\cos\beta$. Use this in the equation for $x(t)$ at the time t of impact,

$$R\cos\beta = (v_0 \cos\theta)t,$$

and solve for the impact time t:

$$t = \frac{R\cos\beta}{v_0\cos\theta}.$$

The y-coordinate of the impact point is $R\sin\beta$. Use this in the equation for $y(t)$, with t the impact time that we just found,

$$R\sin\beta = (v_0\sin\theta)\left(\frac{R\cos\beta}{v_0\cos\theta}\right) - g\frac{\left(\frac{R\cos\beta}{v_0\cos\theta}\right)^2}{2},$$

which simplifies (somewhat) to

$$R\sin\beta = \sin\theta\left(\frac{R\cos\beta}{\cos\theta}\right) - g\frac{\left(\frac{R\cos\beta}{v_0\cos\theta}\right)^2}{2}.$$

Divide each term of the equation by R and isolate the only remaining term with an R on the left hand side:

$$g\frac{R\cos^2\beta}{2v_0^2\cos^2\theta} = \sin\theta\frac{\cos\beta}{\cos\theta} - \sin\beta = \frac{\sin\theta\cos\beta - \cos\theta\sin\beta}{\cos\theta} = \frac{\sin(\theta-\beta)}{\cos\theta}.$$

Solve for R:

$$R = \frac{2v_0^2\cos\theta\sin(\theta-\beta)}{g\cos^2\beta}.$$

4.29

a)   Choose a coordinate system with origin at the frog's launch point, $\hat{\mathbf{i}}$ pointed to the right, and $\hat{\mathbf{j}}$ pointed up. In this coordinate system

$$\begin{aligned} y_0 &= 0\,\text{m} \\ v_{y0} &= v_0, \\ a_y &= -g. \end{aligned}$$

So, the kinematics equations are

$$\begin{aligned} v_y(t) &= v_0 - gt \\ y(t) &= v_0 t - g\frac{t^2}{2}. \end{aligned}$$

When the frog reaches its maximum height, $v_y = 0\,\text{m/s}$. Hence $0\,\text{m/s} = v_0 - gt$, so $t = \dfrac{v_0}{g}$.

When the frog is a maximum height, $y = h$. Hence

$$h = v_0\frac{v_0}{g} - g\frac{\left(\frac{v_0}{g}\right)^2}{2} = \frac{v_0^2}{2}.$$

Therefore $v_0 = \sqrt{2gh}$.

b)   When the frog leaps for maximum horizontal range using the same speed v_0, use the same coordinate system but now

$$
\begin{aligned}
y_0 &= 0 \text{ m} \\
v_{y0} &= v_0 \sin\theta \\
a_y &= -g
\end{aligned}
\qquad\qquad
\begin{aligned}
x_0 &= 0 \text{ m} \\
v_{x0} &= v_0 \cos\theta \\
a_x &= 0 \text{ m/s}^2 \,.
\end{aligned}
$$

So, the kinematics equations are

$$
\begin{aligned}
v_y(t) &= v_0 \sin\theta - gt \\
y(t) &= (v_0 \sin\theta)t - g\frac{t^2}{2}
\end{aligned}
\qquad\qquad
\begin{aligned}
v_x(t) &= v_0 \cos\theta \\
x(t) &= (v_0 \cos\theta)t.
\end{aligned}
$$

Impact occurs where $y = 0$ m , so

$$
0 \text{ m} = (v_0 \sin\theta)t - g\frac{t^2}{2}.
$$

Use the quadratic equation to solve for t. The two roots are

$$
t = 0 \text{ s} \quad \text{and} \quad t = \frac{2v_0 \sin\theta}{g}.
$$

The zero root is the launch time, so we want the nonzero root. The horizontal range R is found by substituting this time into the equation for x:

$$
R = v_0 \cos\theta \frac{2v_0 \sin\theta}{g} = \frac{2v_0^2 \sin\theta \cos\theta}{g} = \frac{v_0^2 \sin 2\theta}{g}.
$$

This a maximum when $2\theta = 90° \implies \theta = 45°$, so

$$
R_{\text{max}} = \frac{v_0^2}{g}.
$$

Substitute the expression for v_0^2 from part a):

$$
R_{\text{max}} = \frac{2gh}{g} = 2h.
$$

c)   The time to the maximum height is half that for the horizontal range, which was found in part b) to be

$$
t = \frac{v_0 \sin\theta}{g}.
$$

The maximum height then is

$$
y_{\text{max}} = v_0 \sin\theta \frac{v_0 \sin\theta}{g} - g\frac{\left(\dfrac{v_0 \sin\theta}{g}\right)^2}{2} = \frac{v_0^2 \sin^2\theta}{2g}.
$$

Substitute $\theta = 45°$ and $v_0^2 = 2gh$ to find $y_{\text{max}} = \dfrac{2gh \sin^2 45°}{2g} = \dfrac{h}{2}.$

4.33   Use a coordinate system with origin at the launch point of the pellet, $\hat{\mathbf{i}}$ pointing to the right, and $\hat{\mathbf{j}}$ pointing up. In this coordinate system

$$
\begin{aligned}
y_0 &= 0 \text{ m} \\
v_{y0} &= v_0 \sin\theta \\
a_y &= -g
\end{aligned}
\qquad\qquad
\begin{aligned}
x_0 &= 0 \text{ m} \\
v_{x0} &= v_0 \cos\theta \\
a_x &= 0 \text{ m/s}^2 \,.
\end{aligned}
$$

So, the kinematics equations are

$$v_y(t) = v_0 \sin\theta - gt$$
$$y(t) = (v_0 \sin\theta)t - g\frac{t^2}{2}$$

$$v_x(t) = v_0 \cos\theta$$
$$x(t) = (v_0 \cos\theta)t.$$

The pellet hits the monkey when the x-coordinate of the pellet is equal to the x-coordinate of the monkey. Denote this value by R. Then

$$R = (v_0 \cos\theta)t,$$

so

$$t = \frac{R}{v_0 \cos\theta}.$$

Substitute this value for t into the first term on the right-hand side of the equation for y:

$$y(t) = v_0 \sin\theta \frac{R}{v_0 \cos\theta} - g\frac{t^2}{2} = r\tan\theta - g\frac{t^2}{2}.$$

Note that $R\tan\theta = h$, where h is the original height of the monkey, so the equation for the y-coordinate of the pellet when it hits the monkey is

$$y(t) = h - g\frac{t^2}{2}.$$

For the vertical motion of the monkey we have $y_0 = h$, $v_{y0} = 0$ m/s, and $a_y = -g$. So the equation for the velocity component is $v_y = -gt$. The equation for the monkey's position is

$$y(t) = h - g\frac{t^2}{2}.$$

Note that the equation for the y-coordinate of the monkey is the same as that for the pellet when it reaches $x = R$. Therefore, the pellet will collide with the falling monkey, since at that point they both have the same x and y coordinates.

4.37   The three velocities are related by the relative velocity addition equation:

$$\vec{v}_{\text{swimmer ground}} = \vec{v}_{\text{swimmer water}} + \vec{v}_{\text{water ground}}.$$

We want the velocity of the swimmer with respect to the ground to be directly across the river, so the three velocities must form a right triangle, as shown below:

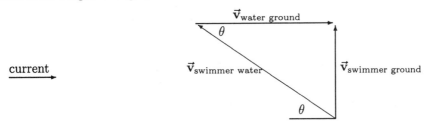

a)   The cosine of the angle θ that $\vec{v}_{\text{swimmer water}}$ makes with the upstream direction is thus

$$\cos\theta = \frac{4.00 \text{ km/h}}{5.00 \text{ km/h}} \quad \Longrightarrow \quad \theta = 36.9°.$$

b)   The speed of the swimmer with respect to the ground is found from the Pythagorean theorem:

$$v_{\text{swimmer ground}} = \sqrt{(5.00 \text{ km/h})^2 - (4.00 \text{ km/h})^2} = 3.00 \text{ km/h}.$$

Convert this to m/s:

$$3.00 \text{ km/h} = (3.00 \text{ km/h}) \left(\frac{10^3 \text{ m}}{\text{km}} \right) \left(\frac{\text{h}}{3600 \text{ s}} \right) = 0.833 \text{ m/s}.$$

The width of the river is 100 (m). The time it takes the swimmer to cross the stream is

$$t = \frac{\text{distance}}{\text{speed}} = \frac{100 \text{ m}}{0.833 \text{ m/s}} = 120 \text{ s}.$$

4.41

a)   The geometry of the situation and a coordinate system are shown below.

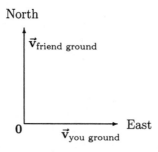

The two velocities are

$$\vec{\mathbf{v}}_{\text{you ground}} = (1.50 \text{ m/s})\hat{\mathbf{i}} \quad \text{and} \quad \vec{\mathbf{v}}_{\text{friend ground}} = (1.00 \text{ m/s})\hat{\mathbf{j}}.$$

The velocity of your friend with respect to you is $\vec{\mathbf{v}}_{\text{friend you}}$. The three velocities are related by the relative velocity addition equation:

$$\vec{\mathbf{v}}_{\text{friend you}} = \vec{\mathbf{v}}_{\text{friend ground}} + \vec{\mathbf{v}}_{\text{ground you}}.$$

The velocity of the ground with respect to you is the opposite of the velocity of you with respect to the ground:

$$\vec{\mathbf{v}}_{\text{ground you}} = -\vec{\mathbf{v}}_{\text{you ground}} = -(1.50 \text{ m/s})\hat{\mathbf{i}}.$$

The relative velocity addition equation thus becomes:

$$
\begin{aligned}
\vec{\mathbf{v}}_{\text{friend you}} &= \vec{\mathbf{v}}_{\text{friend ground}} + \vec{\mathbf{v}}_{\text{ground you}} \\
&= (1.00 \text{ m/s})\hat{\mathbf{j}} - (1.50 \text{ m/s})\hat{\mathbf{i}} \\
&= -(1.50 \text{ m/s})\hat{\mathbf{i}} + (1.00 \text{ m/s})\hat{\mathbf{j}}
\end{aligned}
$$

b)   Note that the arguments in part a) make no mention of the position of either you or your friend. Hence, the velocity of your friend with respect to you does not depend upon where either of you are located.

4.45   The situation and a coordinate system are indicated below:

4.49.

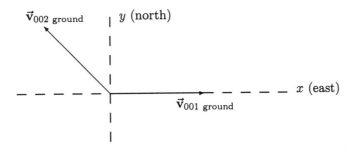

a)   Flight 001 has a velocity with respect to the ground of

$$\vec{v}_{001 \text{ ground}} = (600 \text{ km/h})\hat{\mathbf{i}}.$$

Flight 002 has a velocity with respect to the ground of

$$\vec{v}_{002 \text{ ground}} = (-700\cos(45°) \text{ km/h})\hat{\mathbf{i}} + (700\sin(45°) \text{ km/h})\hat{\mathbf{j}} = (-495 \text{ km/h})\hat{\mathbf{i}} + (495 \text{ km/h})\hat{\mathbf{j}}.$$

The velocity of flight 002 with respect to flight 001 is found from the relative velocity addition equation:

$$\vec{v}_{002 \text{ 001}} = v_{002 \text{ ground}} + \vec{v}_{\text{ground 001}}.$$

But

$$\vec{v}_{\text{ground 001}} = -\vec{v}_{001 \text{ ground}}.$$

Hence, the relative velocity addition equation becomes

$$\vec{v}_{002 \text{ 001}} = (-495 \text{ km/h})\hat{\mathbf{i}} + (495 \text{ km/h})\hat{\mathbf{j}} + (-600 \text{ km/h})\hat{\mathbf{i}} = (-1095 \text{ km/h})\hat{\mathbf{i}} + (495 \text{ km/h})\hat{\mathbf{j}}.$$

b)   The speed is the magnitude of the vector $\vec{v}_{002 \text{ 001}}$:

$$v_{002 \text{ 001}} = \sqrt{(-1095 \text{ km/h})^2 + (495 \text{ km/h})^2} = 1.20 \times 10^3 \text{ km/h}.$$

4.49   Convert revolutions per minute to radians per second:

$$3000 \text{ rev/min} = (3000 \text{ rev/min})\left(\frac{2\pi \text{ rad}}{\text{rev}}\right)\left(\frac{\text{min}}{60 \text{ s}}\right) = 314.2 \text{ rad/s}.$$

4.53   Solve for r:

$$r = \frac{v^2}{a_{\text{centripetal}}} = \frac{(10 \text{ m/s})^2}{9.81 \text{ m/s}} = 10 \text{ m}.$$

4.57   First convert the angular speed to rad/s:

$$\omega = 33 \text{ rev/s} = (33 \text{ rev/s})\left(\frac{2\pi \text{ rad}}{\text{rev}}\right) = 2.1 \times 10^2 \text{ rad/s}.$$

The magnitude of the centripetal acceleration is

$$a_{\text{centripetal}} = r\omega^2 = (10 \text{ km})(2.1 \times 10^2 \text{ rad/s})^2 = (10 \text{ km})\left(\frac{10^3 \text{ m}}{\text{km}}\right)(2.1 \times 10^2 \text{ rad/s})^2 = 4.4 \times 10^8 \text{ m/s}^2.$$

4.61   First convert the speed from km/h to m/s:

$$v = 300 \text{ km/h} = (300 \text{ km/h}) \left(\frac{10^3 \text{ m}}{\text{km}} \right) \left(\frac{\text{h}}{3600 \text{ s}} \right) = 83.3 \text{ m/s}.$$

The magnitude of the centripetal acceleration is not to exceed 4.0% of g. Setting them equal,

$$a_{\text{centripetal}} = 0.040g = 0.39 \text{ m/s}^2 \implies \frac{v^2}{r} = 0.39 \text{ m/s}^2.$$

Hence, at the maximum permissible acceleration, the radius of the curve is

$$r = \frac{(83.383 \text{ m/s})^2}{0.39 \text{ m/s}^2} = 1.8 \times 10^4 \text{ m} = 18 \text{ km}.$$

With a given speed, if the radius decreases, the centripetal acceleration increases, so the curve must have a radius no smaller than the 18 km calculated above.

4.65

a)   Each hand turns through 2π rad in one revolution. Hence the angular speeds are:

$$\text{second hand:} \quad \frac{2\pi \text{ rad}}{60 \text{ s}} \quad \approx 0.105 \text{ rad/s}.$$

$$\text{minute hand:} \quad \frac{2\pi \text{ rad}}{3\,600 \text{ s}} \quad \approx 1.75 \times 10^{-3} \text{ rad/s}.$$

$$\text{hour hand:} \quad \frac{2\pi \text{ rad}}{43\,200 \text{ s}} \quad \approx 1.45 \times 10^{-4} \text{ rad/s}.$$

b)   All such clock hands turn at these angular speeds, so the answers to part a) do not depend upon the radius of the clock hands.

c)   Since the clock hands all turn in the same sense, clockwise (of course!), the angular velocity vectors all point in the same direction, perpendicular to the clock face and *into* the clock.

d)   The ratio of the angular speed of the second hand to that of the minute hand is the ratio of the times for them to make a complete rotation, a factor of 60.

e)   The ratio of the angular speed of the minute hand to that of the hour hand is the ratio of the times for them to make a complete rotation, a factor of 12.

4.69

a)   First convert the angular speeds from (rev/min) to (rad/s):

$$\omega_{z0} = 3.00 \times 10^3 \text{ rev/min} = (3.00 \times 10^3 \text{ rev/min}) \left(\frac{2\pi \text{ rad}}{\text{rev}} \right) \left(\frac{\text{min}}{60 \text{ s}} \right) = 314 \text{ rad/s},$$

$$\omega_z = 4.50 \times 10^3 \text{ rev/min} = (4.50 \times 10^3 \text{ rev/min}) \left(\frac{2\pi \text{ rad}}{\text{rev}} \right) \left(\frac{\text{min}}{60 \text{ s}} \right) = 471 \text{ rad/s}.$$

Solve the rotational kinematics equation

$$\omega_z(t) = \omega_{z0} + \alpha_z t \implies 471 \text{ rad/s} = 314 \text{ rad/s} + \alpha_z(1.20 \text{ s})$$

for α_z:

$$\alpha_z = 131 \text{ rad/s}^2.$$

4.73.

b) Use the equation for the angular position:

$$\theta(t) = \theta_0 + \omega_{z0}t + \alpha_z \frac{t^2}{2}.$$

Choosing $\theta_0 = 0$ rad, this becomes

$$\theta = 0 \text{ rad} + (315 \text{ rad/s})(1.20 \text{ s}) + (131 \text{ rad/s}^2)\frac{(120 \text{ s})^2}{2} = 472 \text{ rad}.$$

Convert this to revolutions:

$$472 \text{ rad} = (472 \text{ rad})\left(\frac{\text{rev}}{2\pi \text{ rad}}\right) = 75.1 \text{ rev}.$$

4.73

a) Convert the final angular speed from (rev/min) to (rad/s):

$$1.00 \times 10^5 \text{ rev/min} = (1.00 \times 10^5 \text{ rev/min})\left(\frac{2\pi \text{ rad}}{\text{rev}}\right)\left(\frac{\text{min}}{60 \text{ s}}\right) = 1.05 \times 10^4 \text{ rad/s}.$$

Choose $\hat{\mathbf{k}}$ to be in the same direction as the angular velocity vector. The angular velocity component at any time is

$$\omega_z(t) = \omega_{z0}t + \alpha_z t^2 \implies 1.05 \times 10^4 \text{ rad/s} = 0 \text{ rad/s} + \alpha_z(420 \text{ s}).$$

Solve for α_z: $\alpha_z = 25.0 \text{ rad/s}^2$.

b) Choosing $\theta_0 = 0$ rad, we have

$$\theta = \theta_0 + \omega_{z0}t + \frac{\alpha_z t^2}{2} \implies \theta = 0 \text{ rad} + (0 \text{ rad/s})(420 \text{ s}) + (25.0 \text{ rad/s}^2)\frac{(420 \text{ s})^2}{2} = 2.21 \times 10^6 \text{ rad}.$$

Convert this angle to revolutions:

$$2.21 \times 10^6 \text{ rad} = (2.21 \times 10^6 \text{ rad})\left(\frac{\text{rev}}{2\pi \text{ rad}}\right) = 3.52 \times 10^5 \text{ rev}.$$

c) The magnitude of the centripetal acceleration is

$$a_{\text{centripetal}} = r\omega^2 = (8.00 \times 10^{-2} \text{ m})(1.05 \times 10^4 \text{ rad/s})^2 = 8.82 \times 10^6 \text{ m/s}^2.$$

Compare this to the magnitude of the local acceleration due to gravity by forming the ratio

$$\frac{a_{\text{centripetal}}}{g} = \frac{8.82 \times 10^6 \text{ m/s}^2}{9.81 \text{ m/s}^2} = 8.99 \times 10^5.$$

Thus, $a_{\text{centripetal}} \approx 900\,000g$!

d) Consider this as a new problem, but use the same coordinate system. The initial angular velocity component is $\omega_{z0} = 1.05 \times 10^4$ rad/s. Hence, when the rotor stops,

$$\omega_z = \omega_{z0} + \alpha_z t \implies 0 \text{ rad/s} = 1.05 \times 10^4 \text{ rad/s} + \alpha_z(240 \text{ s}).$$

Solve for α_z:

$$\alpha_z = -43.8 \text{ rad/s}^2.$$

The angular velocity and angular acceleration vectors are in opposite directions, and the magnitude of the angular acceleration is 43.8 rad/s^2.

4.77

a) The magnitude of the centripetal acceleration is

$$a_{\text{centripetal}} = r\omega^2 \qquad \text{where } \omega = |\omega_z|.$$

Since $\omega_{z0} = 0$ rad/s, then $\omega_z(t) = \alpha_z t$ and $a_{\text{centripetal}} = r(\alpha_z t)^2$. The magnitude of the tangential acceleration is $a_{\text{tangential}} = \alpha r = |\alpha_z| r$. When the magnitude of the centripetal acceleration is equal to the magnitude of the tangential acceleration, we have

$$r\alpha^2 t^2 = \alpha r.$$

Hence $\alpha t^2 = 1$, which implies that $t = \sqrt{\dfrac{1}{\alpha}}$.

b) The tangent of the angle ϕ that the total acceleration makes with the radial direction is

$$\tan\phi = \frac{a_{\text{tangential}}}{a_{\text{centripetal}}}.$$

When the tangential and centripetal accelerations have equal magnitudes, $\tan\phi = 1$, which implies that $\phi = \dfrac{\pi}{4}$ rad.

4.81

a) Let $\hat{\mathbf{k}}$ be in the same direction as the angular velocity vector of the turntable, as determined by the right-hand rule.

b) Since the turntable is turning with a constant angular speed, the time it takes the morsel to traverse half of the circumference (an angle of π rad), is the angle divided by the angular speed: $t = \dfrac{\pi \text{ rad}}{0.500 \text{ rad/s}} = 6.28$ s.

c) Choosing $\theta_0 = 0$ rad, the equation for angular position is $\theta(t) = \alpha_z \dfrac{t^2}{2}$. Millifoot traverses π rad during the 6.28 s. Therefore,

$$\pi \text{ rad} = \alpha_z \frac{(6.28 \text{ s})^2}{2}.$$

Solve for α_z: $\alpha_z = 0.159$ rad/s^2.

d) The angular velocity component of the morsel on the turntable is the same as that of the turntable itself, which is a constant 0.500 rad/s. The angular velocity component of Millifoot is

$$\omega_z(t) = 0 \text{ rad/s} + (0.159 \text{ rad/s}^2)t.$$

When $t = 6.28$ s, this angular velocity component is

$$\omega_z = (0.159 \text{ rad/s}^2)(6.28 \text{ s}) = 1.000 \text{ rad/s}.$$

These angular velocity components are graphed below:

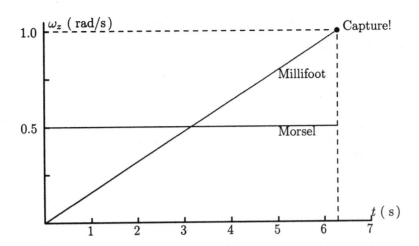

At Capture, the areas under each curve are the same, since at that moment, both Millifoot and Morsel have traveled through the same angle.

4.85

a) The velocity is

$$\vec{v}(t) = \frac{d}{dt}\left(r\cos\theta(t)\,\hat{\mathbf{i}} + r\sin\theta(t)\,\hat{\mathbf{j}}\right) = \left(-r\sin\theta(t)\,\frac{d}{dt}\theta(t)\right)\hat{\mathbf{i}} + \left(r\cos\theta(t)\,\frac{d}{dt}\theta(t)\right)\hat{\mathbf{j}}.$$

The acceleration vector is

$$\begin{aligned}
\vec{\mathbf{a}}(t) &= \frac{d}{dt}\left(-r\sin\theta(t)\,\frac{d}{dt}\theta(t)\,\hat{\mathbf{i}} + r\cos\theta(t)\,\frac{d}{dt}\theta(t)\,\hat{\mathbf{j}}\right) \\
&= \left(-r\sin\theta(t)\,\frac{d^2}{dt^2}\theta(t) - r\cos\theta(t)\left(\frac{d}{dt}\theta(t)\right)^2\right)\hat{\mathbf{i}} + \left(r\cos\theta(t)\,\frac{d^2}{dt^2}\theta(t) - r\sin\theta(t)\left(\frac{d}{dt}\theta(t)\right)^2\right)\hat{\mathbf{j}}.
\end{aligned}$$

b) The acceleration components a_x and a_y are drawn below, along with the unit vectors $\hat{\mathbf{r}}$ and $\hat{\boldsymbol{\theta}}$, so we can see more clearly how the polar coordinates of the acceleration relate to its Cartesian coordinates:

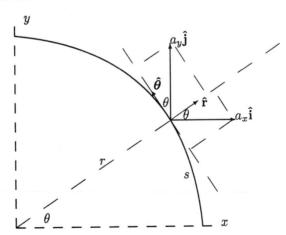

From the sketch, the component $a_x\hat{\mathbf{i}}$ of in the $\hat{\mathbf{r}}$ direction is found by drawing a perpendicular from the tip of $a_x\hat{\mathbf{i}}$ to the line through $\hat{\mathbf{r}}$; this component is $a_x\cos\theta$. Likewise, that of $a_y\hat{\mathbf{j}}$ along $\hat{\mathbf{r}}$ is $a_y\sin\theta$. Hence

$$a_r = a_x\cos\theta + a_y\sin\theta.$$

The component of the acceleration in the tangential direction of increasing θ is found in a similar fashion. From the sketch, the component of $a_x \hat{\mathbf{i}}$ along the $\hat{\boldsymbol{\theta}}$ direction is found by dropping a perpendicular from the tip of $a_x \hat{\mathbf{i}}$ to the line through $\hat{\boldsymbol{\theta}}$; this component is $-a_x \sin \theta$. Likewise, that of $a_y \hat{\mathbf{j}}$ along $\hat{\boldsymbol{\theta}}$ is $a_y \cos \theta$. Hence

$$a_\theta = -a_x \sin \theta + a_y \cos \theta.$$

c)   Let s be the arc length traveled by the particle along the circle, so $s = r\theta$. As the particle moves along the circle, the radial distance remains constant but θ varies with time. Hence

(1) $$\frac{ds}{dt} = r\frac{d\theta}{dt} \quad \text{and} \quad \frac{d^2}{dt^2}s = r\frac{d^2}{dt^2}\theta.$$

But $\dfrac{d^2}{dt^2}\theta$ is the angular acceleration component α_z, so

$$\frac{d^2}{dt^2}s = r\alpha_z.$$

The quantity $r\alpha_z$ is the tangential acceleration component. Therefore, the tangential acceleration component is the same as $\dfrac{d^2}{dt^2}s$. Equation (1) indicates that

$$\frac{ds}{dt} = r\omega_z.$$

But $r|\omega_z|$ is the speed v of the particle. Hence

$$\left|\frac{ds}{dt}\right| = v.$$

The magnitude of the centripetal acceleration is

$$a_{\text{centripetal}} = \frac{v^2}{r} = \frac{\left(\dfrac{ds}{dt}\right)^2}{r}$$

and is directed toward the center of the circle, along $-\hat{\mathbf{r}}$.

Chapter 5

Newton's Laws of Motion

5.1 According to Newton's second law,

$$F_{\text{total}} = ma = (1.25 \times 10^3 \text{ kg})(0.150 \text{ m/s}^2) = 188 \text{ N}.$$

5.5

a) The three forces on the system are $(10.00 \text{ N})\hat{\mathbf{i}}$, $(5.00 \text{ N})\hat{\mathbf{j}}$, and

$$(-15.00 \cos 60.0° \text{ N})\hat{\mathbf{i}} - (15.00 \sin 60.0° \text{ N})\hat{\mathbf{j}} = (-7.50 \text{ N})\hat{\mathbf{i}} - (13.0 \text{ N})\hat{\mathbf{j}}.$$

The total force on the system is the vector sum of these three forces:

$$\vec{\mathbf{F}}_{\text{total}} = (10.00 \text{ N})\hat{\mathbf{i}} + (5.00 \text{ N})\hat{\mathbf{j}} - (7.50 \text{ N})\hat{\mathbf{i}} - (13.0 \text{ N})\hat{\mathbf{j}} = (2.50 \text{ N})\hat{\mathbf{i}} - (8.0 \text{ N})\hat{\mathbf{j}}.$$

This force is sketched below:

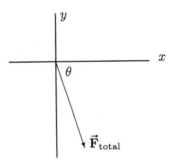

The angle θ that this force makes with $\hat{\mathbf{i}}$ satisfies

$$\tan \theta = \frac{8.0 \text{ N}}{2.50 \text{ N}} = 3.2.$$

Hence, $\theta = 73°$.

b) The magnitude of the total force is

$$F_{\text{total}} = \sqrt{(2.50 \text{ N})^2 + (-8.0 \text{ N})^2} = 8.4 \text{ N}.$$

We also know that $F_{\text{total}} = ma$, so

$$8.4 \text{ N} = (4.00 \text{ kg})a.$$

Therefore, $a = 2.1 \text{ m/s}^2$.

5.9

a) Choose a coordinate system with $\hat{\mathbf{i}}$ along the 500 N force.

b) The total force is

$$F_{\text{total}} = (500\text{ N})\hat{\mathbf{i}} + (250\cos 75.0°\text{ N})\hat{\mathbf{i}} + (250\sin 75.0°\text{ N})\hat{\mathbf{j}} = (565\text{ N})\hat{\mathbf{i}} + (241\text{ N})\hat{\mathbf{j}}.$$

The magnitude of the total force is

$$F_{\text{total}} = \sqrt{(565\text{ N})^2 + (241\text{ N})^2} = 614\text{ N}.$$

c) The magnitude of the total force also is $F_{\text{total}} = ma$, so

$$614\text{ N} = (80\text{ kg})a.$$

Solve this for a: $a = 7.7\text{ m/s}^2$.

d) The acceleration is in the same direction as the total force. A unit vector in the direction of the total force is found by taking the total force and dividing by its magnitude:

$$\frac{1}{F_{\text{total}}}\vec{\mathbf{F}} = \frac{1}{614\text{ N}}\left((565\text{ N})\hat{\mathbf{i}} + (214\text{ N})\hat{\mathbf{j}}\right) = 0.920\hat{\mathbf{i}} + 0.393\hat{\mathbf{j}}.$$

5.13

a) Use the coordinate system shown in the text.

b) Write each force in Cartesian form:

$$\vec{\mathbf{F}}_{\text{public opinion}} = (100\text{ N})\hat{\mathbf{i}}$$

$$\vec{\mathbf{F}}_{\text{normalcy}} = (200\text{ N})\hat{\mathbf{j}}$$

$$\vec{\mathbf{F}}_{\text{lobbying}} = (-400\text{ N})\hat{\mathbf{i}}$$

$$\vec{\mathbf{F}}_{\text{party forces}} = (-300\cos 45.0°\text{ N})\hat{\mathbf{i}} - (300\sin 45.0°\text{ N})\hat{\mathbf{j}}$$

$$= -(212\text{ N})\hat{\mathbf{i}} - (212\text{ N})\hat{\mathbf{j}}$$

$$\vec{\mathbf{F}}_{\text{pacs}} = (150\cos 30.0°\text{ N})\hat{\mathbf{i}} - (150\sin 30.0°\text{ N})\hat{\mathbf{j}}$$

$$= (130\text{ N})\hat{\mathbf{i}} - (75.0\text{ N})\hat{\mathbf{j}}.$$

Use Newton's second law $\vec{\mathbf{F}}_{\text{total}} = m\vec{\mathbf{a}}$ to find the acceleration:

$$(100\ N)\hat{\mathbf{i}} + (200\text{ N})\hat{\mathbf{j}} - (400\text{ N})\hat{\mathbf{i}} - (212\text{ N})\hat{\mathbf{i}} - (212\text{ N})\hat{\mathbf{j}} + (130\text{ N})\hat{\mathbf{i}} - (75.0\text{ N})\hat{\mathbf{j}} = (90\text{ kg})\vec{\mathbf{a}}.$$

Simplifying:

$$(-382\text{ N})\hat{\mathbf{i}} - (87\text{ N})\hat{\mathbf{j}} = (90\text{ kg})\vec{\mathbf{a}}.$$

So $\vec{\mathbf{a}} = (-4.2\text{ m/s}^2)\hat{\mathbf{i}} - (1.0\text{ m/s}^2)\hat{\mathbf{j}}$.

c) The magnitude of the acceleration is

$$a = \sqrt{(-4.2\text{ m/s}^2)^2 + (-1.0\text{ m/s}^2)^2} = 4.3\text{ m/s}^2.$$

5.17 The magnitude of the weight is

$$w = mg \qquad \Longrightarrow \qquad 1.00\text{ N} = m(9.81\text{ m/s}^2).$$

Solve for m: $m = 0.102\text{ kg}$.

5.21

a) The forces on the rock are:

1. the gravitational force of the Earth on the rock (its weight) $\vec{\mathbf{W}}$, equal in magnitude to mg, directed vertically downward; and

2. the normal force $\vec{\mathbf{N}}$ of the bottom of the bucket on the rock, directed vertically upward.

b) Since the rock is in circular motion, the acceleration is the centripetal acceleration directed toward the center of the circle. Hence

$$ma = \frac{mv^2}{r} = \frac{(1.50 \text{ kg})(4.00 \text{ m/s})^2}{1.20 \text{ m}} = 20.0 \frac{\text{kg} \cdot \text{m}}{\text{s}^2} = 20.0 \text{ N}.$$

The magnitude of the total force on the rock is ma, so $F_{\text{total}} = 20.0 \text{ N}$.

c) Choose the coordinate system so that $\hat{\mathbf{j}}$ is pointing straight up. Then Newton's second law becomes

$$-mg\hat{\mathbf{j}} + N\hat{\mathbf{j}} = ma\hat{\mathbf{j}}.$$

Solve for N:

$$N = ma + mg = 20.0 \text{ N} + (1.50 \text{ kg})(9.81 \text{ m/s}^2) = 34.7 \text{ N}.$$

Hence, the normal force of the bottom of the bucket on the rock has magnitude 34.7 N and is directed vertically upward.

d) Find the normal force of the bottom of the bucket on the rock when it is at the top of its circular path. If a nonzero normal force is needed (along with the weight) to provide the centripetal acceleration, the rock will remain in the bucket. When the bucket is at the top of its circular path, the forces on the rock are:

1. the gravitational force of the Earth on the rock (its weight) $\vec{\mathbf{w}}$, equal in magnitude to mg, directed vertically downward;

2. the normal force $\vec{\mathbf{N}}$ of the bottom of the bucket on the rock, directed vertically upward.

Apply Newton's second law to the rock, taking $\hat{\mathbf{j}}$ to be pointing up as before:

$$\vec{\mathbf{F}}_{\text{total}} = m\vec{\mathbf{a}} \qquad \Longrightarrow \qquad -mg\hat{\mathbf{j}} - N'\hat{\mathbf{j}} = -m\frac{v^2}{r}\hat{\mathbf{j}}.$$

Solve for N':

$$N' = \frac{mv^2}{r} - mg = \frac{(1.50 \text{ kg})(4.00 \text{ m/s})^2}{1.20 \text{ m}} - (1.50 \text{ kg})(9.81 \text{ m/s}^2) = 20.0 \text{ N} - 14.7 \text{ N} = 5.3 \text{ N}.$$

Since $N' \neq 0 \text{ N}$ and is positive, the rock will remain in the bucket.

5.25

a) Let θ be the angle that the frictionless plane makes with the horizontal direction. Consider a mass m, placed on the plane, to be the system. The two forces on m are:

1. its weight $\vec{\mathbf{w}}$, directed vertically downward; and

2. the normal force $\vec{\mathbf{N}}$ of the surface on m, oriented perpendicular to the surface.

Use a coordinate system whose x-axis is parallel to the plane and whose y-axis points up, perpendicular to the plane. With this choice,

$$F_{\text{total}, x} = ma_x \quad \Longrightarrow \quad mg\sin\theta = ma_x$$

$$\Longrightarrow \quad g\sin\theta = a_x$$

$$\Longrightarrow \quad \sin\theta = \frac{a_x}{g} = \frac{1.00 \text{ m/s}^2}{9.81 \text{ m/s}^2} = 0.102$$

$$\Longrightarrow \quad \theta = 5.85°.$$

b)   After its release, the forces on the mass are the same as in part a), so the resulting acceleration is also the same.

c)   The magnitude of the total force on the 1.00 kg mass is

$$F_{\text{total}} = ma = (1.00 \text{ kg})(1.00 \text{ m/s}^2) = 1.00 \text{ N}.$$

5.29   When the magnitude of the weight of a mass m hung from the fishing line exceeds 100 N, the cord will break. Hence

$$mg = 100 \text{ N} \implies m = \frac{100 \text{ N}}{9.81 \text{ m/s}^2} = 10.2 \text{ kg}.$$

5.33

a)   The force of the rope on you has a magnitude of 100 N and is directed vertically upward; it is the third-law force pair of the force you exert on the rope.

b)   The second law force diagram includes the following forces:

1. your weight $\vec{\mathbf{w}}$ (the gravitational force of the Earth on you);

2. the normal force $\vec{\mathbf{N}}$ of the floor surface on you, directed vertically upward; and

3. the force of the rope $\vec{\mathbf{T}}$ on you, directed vertically upward.

Note that the length of $\vec{\mathbf{T}} + \vec{\mathbf{N}}$ is equal to the length of $\vec{\mathbf{w}}$.

c)   The corresponding third-law forces are (in the same order):

1. the gravitational force of you on the Earth;

2. the force of you on the floor; and

3. the force of you on the rope.

d)   The magnitude of the gravitational force of the Earth on you is the magnitude of your weight,

$$mg = (70.0 \text{ kg})(9.81 \text{ m/s}^2) = 687 \text{ N}.$$

e)   Apply Newton's second law taking yourself as the system. Let $\hat{\mathbf{j}}$ point straight up. Since your acceleration is zero,

$$\vec{\mathbf{F}}_{\text{total}} = m\vec{\mathbf{a}} \implies \vec{\mathbf{T}} + \vec{\mathbf{w}} + \vec{\mathbf{N}} = m(\vec{\mathbf{0}})$$

$$\implies (100 \text{ N})\hat{\mathbf{j}} - (687 \text{ N})\hat{\mathbf{j}} + \vec{\mathbf{N}} = \vec{\mathbf{0}}$$

$$\implies \vec{\mathbf{N}} = (587 \text{ N})\hat{\mathbf{j}}.$$

The normal force has a magnitude of 587 N.

f)   The scale indicates the force you exert on the floor, which is the third-law counterpart of the normal force of the floor on you. The force has magnitude 587 N. Were the scale to read in kilograms, it would show

$$\frac{587 \text{ N}}{9.81 \text{ m/s}^2} = 59.8 \text{ kg}.$$

5.37

a) Consider each mass to be a separate system. The forces on each are:

 1. its weight; and

 2. the force of the string on it.

Note that since the string is ideal, the force of the string on each mass has the same magnitude T.

 Assume m_2 accelerates downward while m_1 simultaneously goes upward with an acceleration of the same magnitude a. Let $\hat{\mathbf{j}}$ point up. Apply Newton's second law to the motion of each mass. Then, for m_1,

(1) $$F_{y\ \text{total}} = m_1 a_y \implies T - m_1 g = m_1 a \implies T = m_1(g + a).$$

For m_2

(2) $$F_{y\ \text{total}} = m_2 a_y \implies T - m_2 g = m_2(-a).$$

Substitute the expression for T from the last equation in (1) into the last equation in (2):

$$m_1(g + a) - m_2 g = m_2(-a) \implies (m_1 + m_2)a = (m_2 - m_1)g \implies a = \left(\frac{m_2 - m_1}{m_1 + m_2}\right)g.$$

Therefore, the magnitude of the acceleration is

$$a = \left(\frac{|m_2 - m_1|}{m_1 + m_2}\right)g$$

b) From the last equation in (1) in part a) we have.

$$T = m_1(g + a) = m_1 g + m_1 a.$$

Substitute the expression found in part a) for a.

$$T = m_1 g + m_1\left(\frac{m_2 - m_1}{m_1 + m_2}\right)g = \left(\frac{2m_1 m_2}{m_1 + m_2}\right)g.$$

5.41

a) Consider the inmate to be the system. During the descent, the forces acting on the inmate are:

 1. the weight $\vec{\mathbf{w}}$ of the inmate (the gravitational force of the Earth on the inmate). Its magnitude is mg, and it is directed straight down;

 2. the force $\vec{\mathbf{T}}$ of the rope on the inmate, directed straight up.

b) Let $\hat{\mathbf{j}}$ point down. Newton's second law is:

$$F_{y\ \text{total}} = m a_y \implies mg - T = m a_y.$$

When the string has its maximum tension, $T = 500$ N. Hence

$$(60.0\ \text{kg})(9.81\ \text{m/s}^2) - 500\ \text{N} = (60.0\ \text{kg})a_y \implies a_y = (1.48\ \text{m/s}^2).$$

The magnitude of the acceleration is the absolute value of this single component: 1.48 m/s². From Newton's second law, we see that for the magnitude of the acceleration of the inmate to be smaller, the tension would have to be larger. However, 500 N is the largest tension the string can support. Hence, this is the *minimum* value for the acceleration if the string is not to break.

c) Let the inmate begin at rest at the y-axis origin. Then the y-coordinate of the inmate is

$$y(t) = y_0 + v_{y0}t + a_y\frac{t^2}{2} = 0 \text{ m} + (0 \text{ m/s})t + a_y\frac{t^2}{2} = (1.48 \text{ m/s}^2)\frac{t^2}{2}.$$

When the inmate reaches the ground, the y-coordinate is 8.0 m. Hence

$$8.0 \text{ m} = (1.48 \text{ m/s}^2)\frac{t^2}{2}.$$

Solve for t, taking the positive root (since the journey began when $t = 0$ s): $t = 3.29$ s.
   The velocity component of the inmate is

$$v_y(t) = v_{y0} + a_y t = 0 \text{ m/s} + (1.48 \text{ m/s}^2)t = (1.48 \text{ m/s}^2)t.$$

When $t = 3.29$ s, the velocity component is $v_y = (1.48 \text{ m/s}^2)(3.29 \text{ s}) = 4.87$ m/s. The speed of the inmate is the absolute value of this single velocity component, 4.87 m/s.

5.45

a) Consider the elevator car and oaf to be a single system with mass $m = m_{\text{elevator car}} + m_{\text{oaf}}$. The forces on the system are:

   1. its weight $\vec{w}$, of magnitude $(m_{\text{elevator car}} + m_{\text{oaf}})g$, directed downward; and

   2. the force $\vec{T}$ of the cable on the system, directed upward.

Choose $\hat{j}$ to point upward. Then in this case

$$F_{y \text{ total}} = ma_y \implies T - mg = ma_y \implies T = m(g + a_y)$$
$$= (1.00 \times 10^3 \text{ kg} + 100 \text{ kg})(9.81 \text{ m/s}^2 + 1.50 \text{ m/s}^2)$$
$$= 1.24 \times 10^4 \text{ N} = 12.4 \text{ kN}.$$

b) Now consider the oaf to be the system. The forces on the oaf are:

   1. its weight $\vec{w}'$, of magnitude $m_{\text{oaf}}g$, directed downward: and

   2. the normal force $\vec{N}$ of the floor on it, directed upward.

Let $\hat{j}$ point up. In this case

$$F_{y \text{ total}} = m_{\text{oaf}}a_y \implies N - m_{\text{oaf}}g = m_{\text{oaf}}a_y \implies N = m_{\text{oaf}}(g + a_y)$$
$$= (100 \text{ kg})(9.81 \text{ m/s}^2 + 1.50 \text{ m/s}^2)$$
$$= 1.13 \times 10^3 \text{ N} = 1.13 \text{ kN}.$$

The magnitude of the normal force is greater than the weight of the oaf so the total force on the oaf is in the direction of its upward acceleration.

5.49

a) Consider the Dean to be the system (in more ways than one!). There are two forces acting on the Dean:

   1. the weight $\vec{w}$ of the Dean (the gravitational force of the Earth on the Dean), of magnitude mg, directed downward; and

   2. the normal force $\vec{N}$ of the floor of the elevator car on the Dean, pointing upward.

b) The gravitational force of the Earth on the Dean has magnitude $mg = (80.0 \text{ kg})(9.81 \text{ m/s}^2) = 785$ N.

c)   When the elevator is at rest, the acceleration of the Dean is zero, so the total force on the Dean also must be zero, according to Newton's second law. Let $\hat{\jmath}$ point upward. Then

$$F_{y \text{ total}} = ma_y \implies N + (-mg) = m(0 \text{ m/s}^2) \implies N = mg = 785 \text{ N}.$$

d)   The second law force diagram is unchanged, except that the magnitude of $\vec{N}$ is greater than $\vec{w}$.

e)   The magnitude of the weight of the Dean is unchanged by the motion of the elevator car: $mg = 785 \text{ N}$.

f)   During the upward acceleration of the elevator car,

$$F_{y \text{ total}} = ma_y \implies N + (-mg) = ma_y \implies N = m(g + a_y) = (80.0 \text{ kg})(9.81 \text{ m/s}^2 + 1.00 \text{ m/s}^2) = 865 \text{ N}.$$

g)   The scale displays the magnitude of the force of the Dean on the scale. This force is the Newton's third-law pair to the force of the scale on the Dean, which is the normal force of the surface on the Dean. Thus, the scale displays 865 N. The magnitude of the normal force of the surface on the Dean must be greater than the magnitude of the weight in order for the total force to be in the upward direction and cause the Dean to accelerate.

h)   The second law force diagram remains unchanged, except that now the normal force has a smaller magnitude than that of the weight of the Dean.

i)   The magnitude of the gravitational force of the Earth on the Dean is not affected by the motion of the elevator car: the weight remains of magnitude 785 N.

j)   Apply Newton's second law once again to the Dean system, realizing that now the acceleration component a_y is -1.00 m/s^2:

$$
\begin{aligned}
F_{y \text{ total}} = ma_y \implies N + (-mg) = ma_y \implies N &= m(g + a_y) \\
&= (80.0 \text{ kg})(9.81 \text{ m/s}^2 - 1.00 \text{ m/s}^2) \\
&= 705 \text{ N}.
\end{aligned}
$$

k)   As in part g), the scale displays the magnitude of the force of the Dean on its surface, which is the third law counterpart to the force of the surface on the Dean. The scale displays 705 N.

5.53   The forces on the anchor are:

1. its weight $\vec{w}$;

2. the normal force $\vec{N}$ of the surface on the anchor;

3. the applied horizontal force $\vec{F}$, of magnitude 350 N; and

4. the maximum static force of friction $\vec{f}_{s \text{ max}}$, since the system is on the verge of slipping.

Since the mass is not accelerating, the total force on the system must be zero. Choose a coordinate system with origin on the mass, $\hat{\imath}$ pointing to the right, and $\hat{\jmath}$ pointing up. In this case

x direction	y direction
$F_{x \text{ total}} = ma_x \implies 350 \text{ N} - f_{s \text{ max}} = 0 \text{ N}$	$F_{y \text{ total}} = ma_y \implies N - mg = 0 \text{ N}$
$\implies f_{s \text{ max}} = 350 \text{ N}$	$\implies N = mg.$

Since the system is on the verge of slipping, we have

$$f_{s \text{ max}} = \mu_s N = \mu_s mg \implies 350 \text{ N} = \mu_s(50 \text{ kg})(9.81 \text{ m/s}^2) \implies \mu_s = 0.71.$$

5.57   Consider the skier to be the system. The forces on the skier are

1. the weight $\vec{w}$ of the skier;

2. the normal force $\vec{N}$ of the slope on the skier; and

3. the kinetic force of friction $\vec{f}_k$ on the skier.

   Choose a coordinate system so that $\hat{i}$ is parallel to the slope in the direction of motion of the skier, and $\hat{j}$ is perpendicular to the surface of the slope, not quite straight up, but outwards from the slope. Apply Newton's second law to the skier along each coordinate axis: Since the skier is moving at constant velocity down the slope, the acceleration of the skier is zero. Let $\theta = 5°$, the angle of the slope. Then

x-direction

$$F_{x\text{ total}} = ma_x \implies mg\sin\theta - f_k = m(0\text{ m/s}^2)$$
$$\implies f_k = mg\sin\theta.$$

y-direction

$$F_{y\text{ total}} = ma_y \implies N - mg\cos\theta = m(0\text{ m/s}^2)$$
$$\implies N = mg\cos\theta.$$

The magnitude of the kinetic force of friction is related to the magnitude of the normal force by

$$f_k = \mu_k N \implies mg\sin\theta = \mu_k mg\cos\theta \implies \mu_k = \tan\theta = \tan 5° = 0.09.$$

5.61

a)   The forces acting on m are:

1. its weight $\vec{w}$, directed downward; and

2. the force $\vec{T}$ of the cord on m, directed upward along the cord.

   The forces on the 50 (kg) mass are:

1. its weight $\vec{W}'$, directed downward;

2. the normal force $\vec{N}$ of the surface on this mass, directed perpendicular to the surface;

3. the force $\vec{T}$ of the cord on the 50 kg mass, directed horizontally to the right along the cord; and

4. the static force of friction $\vec{f}_s$ on the 50 kg mass, directed so as to oppose slippage, i.e., parallel to the surface and to the left.

b)   Choose a coordinate system with $\hat{i}$ pointing to the right and $\hat{j}$ pointing up. When m has its maximum value, the 50 kg mass is ready to slip. Under these conditions, the static force of friction on the 50 kg mass has its greatest magnitude: $f_{s\text{ max}} = \mu_s N$. Apply Newton's second law to the 50 kg mass system in each direction, noting that the acceleration of the system is zero:

x direction

$$F_{x\text{ total}} = (50\text{ kg})a_x$$
$$\implies T - f_{s\text{ max}} = (50\text{ kg})(0\text{ m/s}^2)$$
$$\implies T = f_{s\text{ max}}$$

y direction

$$F_{y\text{ total}} = (50\text{ kg})a_y$$
$$\implies N - (50\text{ kg})g = (50\text{ kg})(0\text{ m/s}^2)$$
$$\implies N = (50\text{ kg})(9.81\text{ m/s}^2) = 491\text{ N}.$$

   Hence, we have

(1) $$T = \mu_s N = (0.20)(491\text{ N}) = 98\text{ N}.$$

Apply Newton's second law to the m system:

$$F_{y\ total} = ma_y \implies T - mg = m(0\ \text{m/s}^2) \implies T = mg.$$

Substitute the value $T = 98\ \text{N}$ from equation (1):

$$98\ \text{N} = mg = m(9.81\ \text{m/s}^2) \implies m = 10\ \text{kg}.$$

c) If m is only half the value in part b), then $m = 5\ \text{kg}$. The magnitude of the tension in the cord then is

$$T = mg = (5\ \text{kg})(9.81\ \text{m/s}^2) = 49\ \text{N}.$$

Consider the 50 kg system and its second law force diagram. Since the system has zero acceleration, Newton's second law for it becomes:

$$F_{x\ total} = (50\ \text{kg})a_x \implies T - f_s = (50\ \text{kg})(0\ \text{m/s}^2) \implies T = f_s.$$

The magnitude of the tension in the cord is equal to the magnitude of the static force of friction on the 50 kg mass system. Hence the magnitude of the static force of friction on the 50 kg mass is 49 N.

5.65

a) If the hanging mass is zero, the force of the cord on you is zero. In this case, the forces acting on you are:

1. your weight $\vec{\mathbf{w}}$, acting downward;

2. the normal force $\vec{\mathbf{N}}$ of the surface on you, directed perpendicular to the surface; and

3. a force of friction $\vec{\mathbf{f}}$, directed parallel to the surface. (If you do not slip, this is a static force of friction.)

You will slip down the plane if the component of your weight down the plane exceeds the maximum magnitude of the static force of friction up the plane. The component of your weight down the plane is

$$m'g \sin\theta = (70\ \text{kg})(9.81\ \text{m/s}^2)\sin 40° = 4.4 \times 10^2\ \text{N}.$$

There is zero acceleration of the system perpendicular to the plane. Choosing a coordinate system with $\hat{\mathbf{j}}$ in this direction we have

$$F_{y\ total} = m'a_y \implies N - m'g\cos\theta = m'(0\ \text{m/s}^2) \implies N = m'g\cos\theta.$$

The maximum magnitude of the static force of friction is

$$f_{s\ max} = \mu_s N = \mu_s m'g\cos\theta = 0.40(70\ \text{kg})(9.81\ \text{m/s}^2)\cos 40° = 2.1 \times 10^2\ \text{N}.$$

Since the component of your weight down the incline is greater than the maximum magnitude of the static force of friction, you will slide down the plane.

b) Your weight $\vec{\mathbf{w}}$ and the normal force of the inclined surface on you are unchanged. Since we need to determine the mass m that enables you to accelerate up the inclined plane, the force of friction is the kinetic force of friction directed opposite to your velocity. There is an additional force acting on you too, the force of the cord.

Two forces act on the hanging mass m: its weight, which points down, and the tension of the cord, which points up. Choose $\hat{\mathbf{j}}$ to point down. In this case we have

$$mg - T = ma \implies T = m(g - a).$$

Now apply Newton's second law to the mass m' on the inclined plane, choosing a coordinate system with $\hat{\mathbf{i}}$ along the cord and $\hat{\mathbf{j}}$ along $\vec{\mathbf{N}}$. In this case,

x direction	y direction

$$F_{x \text{ total}} = ma_x \qquad\qquad\qquad F_{y \text{ total}} = ma_y$$

$$\implies T - m'g\sin\theta - f_k = m'a \qquad\qquad \implies N - m'g\cos\theta = m'(0 \text{ m/s}^2)$$

$$\implies T - m'g\sin\theta - \mu_k N = m'a \qquad\qquad \implies N = m'g\cos\theta.$$

Substitute for T and N in the x-direction equation:

$$m(g - a) - m'g\sin\theta - \mu_k m'g\cos\theta = m'a$$

Solve for m:

$$m = m'\left(\frac{g\sin\theta + \mu_k g\cos\theta + a}{g - a}\right)$$

$$= 70 \text{ kg} \left(\frac{(9.81 \text{ m/s}^2)\sin 40° + 0.35(9.81 \text{ m/s}^2)\cos 40° + 1.50 \text{ m/s}^2}{9.81 \text{ m/s}^2 - 1.50 \text{ m/s}^2}\right)$$

$$= 88 \text{ kg}$$

c)   From b)

$$T = m(g - a) = (88 \text{ kg})(9.81 \text{ m/s}^2 - 1.50 \text{ m/s}^2) = 7.3 \times 10^2 \text{ N}.$$

5.69   The forces on the crate when it is ready to slip are:

1. the weight $\vec{w}$ of the crate, directed vertically downward;

2. the normal force $\vec{N}$ of the surface on the crate; and

3. the maximum value of the static force of friction $\vec{f}_{s \text{ max}}$ on the crate, acting in a direction to oppose slippage.

If the truck accelerates in one direction, the static force of friction will be in the same direction, since the crate will tend to slip in the opposite direction.

Choose a coordinate system with $\hat{i}$ in the direction of motion of the pickup truck. Choose $\hat{j}$ in the direction of $\vec{N}$, pointing upwards from the bed of the pickup. Then

x direction	y direction

$$F_{x \text{ total}} = ma_x \implies f_{s \text{ max}} = ma_x \qquad\qquad F_{y \text{ total}} = ma_y \implies N - mg = m(0 \text{ m/s}^2)$$

$$\implies \mu_s N = ma_x \qquad\qquad\qquad\qquad\qquad \implies N = mg$$

Hence

$$\mu_s mg = ma_x \implies a_x = \mu_s g = 0.35(9.81 \text{ m/s}^2) = 3.4 \text{ m/s}^2$$

5.73

a)   The direction of the total force on you is the same direction as your acceleration. Since you are in circular motion at constant angular speed, your acceleration is the centripetal acceleration toward the center of the merry-go-round. Therefore, the total force also is toward the center of the circular path.

b)   The speed v is

$$v = r\omega = (2.0 \text{ m})(1.00 \text{ rad/s}) = 2.0 \text{ m/s}.$$

c) The forces acting on you are:

1. your weight $\vec{\mathbf{w}}$, directed vertically downward;

2. the normal force $\vec{\mathbf{N}}$ of the surface on you;

3. the static force of friction $\vec{\mathbf{f}}_s$ on you, acting in a direction to prevent slippage. This force points toward the center of the circle.

Choose a coordinate system with $\hat{\mathbf{i}}$ pointed towards the center of the circle and $\hat{\mathbf{j}}$ parallel with $\vec{\mathbf{N}}$, directed perpendicularly upwards from the surface. Apply Newton's second law to each coordinate direction:

x direction y direction

$$F_{x\ \text{total}} = ma_x \implies f_s = m\frac{v^2}{r} \qquad\qquad F_{y\ \text{total}} = ma_y \implies N - mg = m(0\ \text{m/s}^2) \implies N = mg$$

When you are ready to slip, the static force of friction has its maximum magnitude, so

$$f_{s\ \text{max}} = m\frac{v^2}{r} \implies \mu_s N = m\frac{v^2}{r} \implies \mu_s mg = m\frac{v^2}{r}.$$

But $v = r\omega$, so

$$\mu_s mg = m\frac{(r\omega)^2}{r} \implies r = \frac{\mu_s g}{\omega^2} = \frac{0.80(9.81\ \text{m/s}^2)}{(1.00\ \text{rad/s})^2} = 7.8\ \text{m}.$$

d) An observer on the ground would see the projectile follow a parabolic path. The initial velocity of the projectile has a horizontal component equal to 2.0 m/s and a vertical component equal to the speed with which the ball was tossed vertically. From the perspective of the person on the merry-go-round, the ball would go up and down while curving to the right (if the spin of the merry-go-round is counterclockwise when seen from above).

5.77

a) See part b) below.

b) The forces on the puck are:

1. its weight $\vec{\mathbf{w}}$, directed downward;

2. the normal force $\vec{\mathbf{N}}$ of the surface on the puck, directed upward; and

3. the kinetic force of friction $\vec{\mathbf{f}}_k$ on the puck, directed opposite to the velocity of the puck.

Choose a coordinate system with $\hat{\mathbf{i}}$ in the direction of motion of the puck, $\hat{\mathbf{j}}$ in the same direction as $\vec{\mathbf{N}}$ (pointed up), and origin at the initial position of the puck.

c) Apply Newton's second law to the horizontal and vertical directions:

x direction y direction

$$F_{x\ \text{total}} = ma_x \quad\implies\quad -f_k = ma_x \qquad\qquad F_{y\ \text{total}} = may \quad\implies\quad N - mg = m(0\ \text{m/s}^2)$$

$$\implies -\mu_k N = ma_x \qquad\qquad\qquad\qquad\qquad \implies N = mg.$$

Substitute the y-direction equation for N into the last x-direction equation:

$$-\mu_k mg = ma_x \implies a_x = \mu_x g.$$

Now use the kinematic equations for motion with a constant acceleration to determine the velocity component of the puck:

$$v_x(t) = v_{x0} + a_x t = v_0 - \mu_k gt.$$

d)   When the puck comes to rest, $v_x(t) = 0$ m/s, so

$$0 \text{ m/s} = v_0 - \mu_k g t \implies t = \frac{v_0}{\mu_k g}.$$

e)   Since the puck's initial position is the origin:

$$x(t) = x_0 + v_{x0}t + a_x \frac{t^2}{2} = v_0 t - \mu_k g \frac{t^2}{2}.$$

f)   The puck comes to rest when $t = \dfrac{v_0}{\mu_k g}$, so

$$x = v_0 \frac{v_0}{\mu_k g} - \mu_k g \frac{\left(\frac{v_0}{\mu_k g}\right)^2}{2} = \frac{v_0^2}{2\mu_k g}.$$

The distance traveled by the puck is

$$x - x_0 = x - 0 \text{ m} = \frac{v_0^2}{2\mu_k g}.$$

g)   Use the given data to evaluate x:

$$x = \frac{(25 \text{ m/s})^2}{2(0.050)(9.81 \text{ m/s}^2)} = 6.4 \times 10^2 \text{ m}.$$

5.81

a)   The forces on the tire while it is skidding on the pavement are:

1. its weight $\vec{w}$, directed downward;

2. the normal force $\vec{N}$ of the surface on the tire, directed upward; and

3. the force of kinetic friction $\vec{f_k}$ on the tire, directed opposite to velocity of the tire at the point of contact. Since the tire particles at the surface are moving towards the rear of the car, the force of kinetic friction on the tire is directed the same way as the tire as a whole is moving.

b)   The only force in the horizontal direction is the force of kinetic friction; hence it is this force that accelerates the tire.

c)   Write Newton's second law for the tire. Choose a coordinate system with $\hat{\imath}$ in the direction that the car is headed, and $\hat{\jmath}$ pointed up, in the same direction as $\vec{N}$. In this case

x direction   y direction

$$F_{x \text{ total}} = ma_x \implies f_k = ma_x \qquad\qquad F_{y \text{ total}} = ma_y \implies N - mg = m(0 \text{ m/s}^2)$$

$$\implies \mu_k N = ma_x. \qquad\qquad\qquad\qquad\qquad \implies N = mg.$$

Substitute for N in the x equation and solve for a_x:

$$\mu_k mg = ma_x \implies a_x = \mu_k g.$$

5.85   In the expressions for the static coefficient of friction derived in parts c) and f) of the previous problem, the condition $m_2 < m_1 \sin\theta$ must be satisfied to have a positive value for μ_s (which is intrinsically positive since it is the ratio of the magnitudes of two forces). Hence $\dfrac{m_2}{m_1} < \sin\theta$. Since the sine of an angle always is less than or equal to 1, this implies that we also have $\dfrac{m_2}{m_1} < 1$.

a)   When $\phi = \theta$, the expression for the mass ratio becomes

$$\frac{m_2}{m_1} = \frac{\tan\theta + \tan\theta}{\sec\theta + \sec\theta} = \frac{\tan\theta}{\sec\theta} = \frac{\left(\dfrac{\sin\theta}{\cos\theta}\right)}{\left(\dfrac{1}{\cos\theta}\right)} = \sin\theta.$$

b)   Rewrite the mass ratio in terms of sines and cosines.

$$\frac{m_2}{m_1} = \frac{tan\theta + \tan\phi}{\sec\theta + \sec\phi} = \frac{\dfrac{\sin\theta}{\cos\theta} + \dfrac{\sin\phi}{\cos\phi}}{\dfrac{1}{\cos\theta} + \dfrac{1}{\cos\phi}} = \frac{\sin\theta\cos\phi + \sin\phi\cos\theta}{\cos\theta + \cos\phi}.$$

When $\phi = 90°$, the mass ratio becomes

$$\frac{m_2}{m_1} = \frac{\cos\theta}{\cos\theta} = 1.$$

c)   The mass ratio must be between $\sin\theta$ and 1. Hence, the region of possible values is as depicted below.

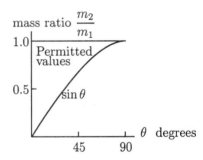

5.89   The magnitude of the centripetal acceleration is

$$a_{\text{centripetal}} = r\omega^2$$

where r is the radius of the circular motion and ω is the angular velocity.

The angular speed of the rotation of Jupiter is

$$\omega = \left(\frac{2\pi \text{ rad}}{9.83 \text{ h}}\right)\left(\frac{\text{h}}{3600 \text{ s}}\right) = 1.78 \times 10^{-4} \text{ rad/s}.$$

The equatorial radius of Jupiter is 7.19×10^7 m. Hence the magnitude of the centripetal acceleration on the equator of Jupiter is

$$a_{\text{centripetal}} = (7.19 \times 10^7 \text{ m})(1.78 \times 10^{-4} \text{ rad/s})^2 = 2.28 \text{ m/s}^2.$$

The ratio between the magnitude of the centripetal acceleration on the equator and the magnitude of the local acceleration due to gravity is

$$\frac{a_{\text{centripetal}}}{a_{\text{gravity}}} = \frac{2.28 \text{ m/s}^2}{24.5 \text{ m/s}^2} = 0.0931 = 9.31\%.$$

Performing the same calculation for Earth we have

$$\omega_{\text{E}} = \left(\frac{2\pi \text{ rad}}{24 \text{ h}}\right)\left(\frac{\text{h}}{3600 \text{ s}}\right) = 7.27 \times 10^{-5} \text{ rad/s},$$

so

$$a_{\text{E centripetal}} = r_E \omega_E^2 = (6.37 \times 10^6 \text{ m})(7.27 \times 10^{-5} \text{ rad/s})^2 = .034 \text{ m/s}^2 \, .$$

Therefore for the Earth, this ratio is

$$\frac{a_{\text{E centripetal}}}{a_{\text{E gravity}}} = \frac{.034 \text{ m/s}^2}{9.81 \text{ m/s}^2} = 0.0035 = 0.35\%.$$

The relatively large value of this ratio on Jupiter, compared with the similar ratio for the Earth means that ignoring the effects of pseudo forces on Jupiter produces less accurate predictions than ignoring them on Earth.

Chapter 6

The Gravitational Force and the Gravitational Field

6.1   The magnitude of the gravitational force is

$$F = \frac{GMm}{r^2} = \frac{(6.67 \times 10^{-11} \text{ N} \cdot \text{m}^2/\text{kg}^2)(2.50 \text{ kg})(2.50 \text{ kg})}{(1.50 \text{ m})^2} = 1.85 \times 10^{-10} \text{ N}.$$

6.5

a)   Assume that the gravitational force of the Sun on the Earth is the only force acting on it. Then the acceleration of the Earth is the centripetal acceleration associated with circular motion. Let M be the mass of the Sun and m be that of the Earth; let r be the radius of the circular orbit of the Earth. In this case

$$F_{\text{total}} = ma \implies \frac{GMm}{r^2} = \frac{mv^2}{r}.$$

Solve for v:

$$v = \sqrt{\frac{GM}{r}} = \sqrt{\frac{(6.67 \times 10^{-11} \text{ N} \cdot \text{m}^2/\text{kg}^2)(1.99 \times 10^{30} \text{ kg})}{1.496 \times 10^{11} \text{ m}}} = 2.98 \times 10^4 \text{ m/s} = 29.8 \text{ km/s}.$$

b)   The orbital speed (see part a) above) is independent of the mass of the orbiting particle, even though it does depend upon the mass of the Sun. Hence, nothing would happen to the orbital speed of each half.

6.9

a)   The magnitude of the gravitational force of the 4.00 kg sphere on the 0.400 kg sphere is

$$F = \frac{GMm}{r^2} = \frac{(6.67 \times 10^{-11} \text{ N} \cdot \text{m}^2/\text{kg}^2)(4.00 \text{ kg})(0.400 \text{ kg})}{(0.101 \text{ m})^2} = 1.05 \times 10^{-8} \text{ N}.$$

b)   The magnitude of the gravitational force of the 0.400 kg sphere on the 4.00 kg sphere is the same as that in part a), since they are a third law force pair.

c)   Find the magnitude of the acceleration of each sphere by applying Newton's second law to each: For the 4.00 kg sphere:

$$F_{\text{total}} = ma \implies 1.05 \times 10^{-8} \text{ N} = (4.00 \text{ kg})a \implies a = 2.63 \times 10^{-9} \text{ m/s}^2.$$

For the 0.400 kg sphere:

$$F_{\text{total}} = m'a' \implies 1.05 \times 10^{-8} \text{ N} = (0.400 \text{ kg})a' \implies a' = 2.63 \times 10^{-8} \text{ m/s}^2.$$

6.13

a) The mass M of the asteroid is the product of its average density and its volume: $M = \rho \left(\frac{4}{3} \pi R^3 \right)$.

Let the gravitational force be the only force acting on a mass m close to the surface of the asteroid. Apply Newton's second law to m, using the magnitudes of the total force and the acceleration:

$$F_{\text{total}} = ma \implies \frac{GMm}{R^2} = ma \implies a = \frac{GM}{R^2} \implies a = \frac{G\rho \left(\frac{4}{3} \pi R^3 \right)}{R^2} \implies a = \frac{4}{3} \pi G R \rho.$$

b) Use the given data in the equation for a from part a):

$$a = \frac{4}{3} \pi (6.67 \times 10^{-11} \text{ N} \cdot \text{m}^2 / \text{kg}^2)(2.26 \times 10^5 \text{ m})(4.5 \times 10^3 \text{ kg/m}^3) = 0.16 \text{ m/s}^2 \, .$$

c) First find the initial vertical speed needed to jump 2.0 m on the surface of the Earth. Use the 1-dimensional kinematic equations for a constant acceleration, and take $\hat{\jmath}$ to be pointing upward.

$$v_y(t) = v_{y0} + a_y t, \qquad \text{and} \qquad y(t) = y_0 + v_{y0} t + a_y \frac{t^2}{2}.$$

When at maximum height, the velocity component is zero, so from the velocity equation,

$$0 \text{ m/s} = v_{y0} - gt, \implies t = \frac{v_{y0}}{g}.$$

Use this value of t in the position equation,

$$2.0 \text{ m} = 0 \text{ m} + v_{y0} t - g \frac{t^2}{2} = v_{y0} \left(\frac{v_{y0}}{g} \right) - g \frac{\left(\frac{v_{y0}}{g} \right)^2}{2} \implies v_{y0} = 6.3 \text{ m/s}.$$

Assume you can jump with the same initial vertical velocity component on the asteroid Camilla, where the magnitude of the surface acceleration due to gravity is only 0.16 m/s^2. Use the kinematic equation for a constant acceleration, and again take $\hat{\jmath}$ to be pointing upward. Then

$$v_y(t) = v_{y0} + a_y t = 6.3 \text{ m/s} - (0.16 \text{ m/s}^2)t.$$

When at maximum height, the vertical velocity component is zero, so

$$0 \text{ m/s} = 6.3 \text{ m/s} - (0.16 \text{ m/s}^2)t \implies t = 39 \text{ s}.$$

Use the equation for y to determine the maximum height:

$$y(t) = y_0 + v_{y0} t + a_y \frac{t^2}{2} = 0(m) + (6.3 \text{ m/s})(39 \text{ s}) - (0.16 \text{ m/s}^2)\frac{(39 \text{ s})^2}{2} = 1.2 \times 10^2 \text{ m}.$$

Since it took 39 s to reach maximum height from the asteroid surface, it will take another 39 s for you to descend to the surface. Your total time of flight, therefore, is 78 s.

6.17

a) The separation of the centers of the spheres is twice the radius of either sphere, so the magnitude of the gravitational force that each sphere exerts on the other is

$$F = \frac{Gmm}{(2R)^2} = \frac{Gm^2}{4R^2}.$$

The mass of each sphere is its density times its volume,

$$m = \rho V = \rho \left(\frac{4}{3} \pi R^3 \right).$$

Substitute this value of m in the expression for the magnitude of their mutual gravitational force:

$$F = \frac{G \left(\rho \left(\frac{4}{3} \pi R^3 \right) \right)^2}{4R^2} = \frac{4G \rho^2 \pi^2 R^6}{9R^2} \implies \rho = \sqrt{\frac{9F}{4\pi^2 G R^4}}.$$

b)   If $F = 1.00$ N and $R = 1.00 \times 10^{-2}$ m, then the requisite density is

$$\sqrt{\frac{9(1.00 \text{ N})}{4\pi^2 (6.67 \times 10^{-11} \text{ N} \cdot \text{m}^2/\text{kg}^2)(1.00 \times 10^{-2} \text{ m})^4}} = 5.85 \times 10^8 \text{ kg/m}^3.$$

No element has a density this great.

c)   From part a), $\rho = \sqrt{\frac{9F}{4\pi^2 G R^4}}$, so solving for R, $R = \sqrt[4]{\frac{9F}{4\pi^2 G \rho^2}}$. Substitute 1.00 N for F and the density of osmium (22.5×10^3) kg/m^3 for ρ,

$$R = \sqrt[4]{\frac{9(1.00 \text{ N})}{4\pi^2 (6.67 \times 10^{-11} \text{ N} \cdot \text{m}^2/\text{kg}^2)(22.5 \times 10^3 \text{ kg/m}^3)^2}} = 1.61 \text{ m}.$$

The mass of each sphere is its density times its volume:

$$m = \rho V = \rho \left(\frac{4}{3} \pi R^3 \right) = (22.5 \times 10^3 \text{ kg/m}^3) \frac{4}{3} \pi (1.61 \text{ m})^3 = 3.93 \times 10^5 \text{ kg}.$$

6.21   Consider the mass m in free-fall near the surface of Mars. Let R be the radius of Mars and M_{Mars} its mass. The only force on m is the gravitational force of Mars on it. Therefore

$$F_{\text{total}} = ma \implies \frac{GM_{\text{Mars}}m}{R^2} = ma \implies M_{\text{Mars}} = \frac{aR^2}{G} = \frac{(3.776 \text{ m/s}^2)(3.37 \times 10^6 \text{ m})^2}{6.67 \times 10^{-11} \text{ N} \cdot \text{m}^2/\text{kg}^2} = 6.43 \times 10^{23} \text{ kg}.$$

Then the ratio

$$\frac{M_{\text{Mars}}}{M_{\text{Earth}}} = \frac{6.43 \times 10^{23} \text{ kg}}{5.98 \times 10^{24} \text{ kg}} = 0.108.$$

The mass of Mars is only about 11% the mass of Earth.

6.25

a)   The distance between the center of the ellipse and either focus is c. Hence the distance between the two foci is $2c$. The definition of the eccentricity of an ellipse is $\epsilon = \frac{c}{a}$, where a is the length of the semimajor axis. Hence the distance between the two foci is

$$2c = 2\epsilon a = 2(0.0167)(1.496 \times 10^8 \text{ km}) = 5.00 \times 10^6 \text{ km}.$$

b)   The distance between either focus and the center of the orbit is half the distance between the foci, or 2.50×10^6 km. The Sun is at one focus of the ellipse. The radius of the Sun is less than 2.50×10^6, so the geometric center of the elliptical orbit of the Earth lies outside the Sun.

c)   The perihelion distance is

$$a - c = a - a\epsilon = a(1 - \epsilon) = (1.496 \times 10^8 \text{ km})(1 - 0.0167) = 1.471 \times 10^8 \text{ km}.$$

The aphelion distance is

$$a + c = a + a\epsilon = a(1 + \epsilon) = (1.496 \times 10^8 \text{ km})(1 + 0.0167) = 1.521 \times 10^8 \text{ km}.$$

d)   The separation of the foci in the model is

$$2c = 2\epsilon a = 2(0.0167)(10.0 \text{ cm}) = 0.334 \text{ cm}.$$

6.29   Let D be the diameter of the Moon. The angle (in radians) subtended by the Moon when it is at its perigee distance is $\theta_{\text{perigee}} = \dfrac{D}{r_{\text{perigee}}}$ (we're actually using a small angle approximation here). The angle (in radians) subtended by the Moon when it is at its apogee distance is $\theta_{\text{apogee}} = \dfrac{D}{r_{\text{apogee}}}$. The ratio of these two angles is

$$\frac{\theta_{\text{perigee}}}{\theta_{\text{apogee}}} = \frac{\dfrac{D}{r_{\text{perigee}}}}{\dfrac{D}{r_{\text{apogee}}}} = \frac{r_{\text{apogee}}}{r_{\text{perigee}}} = \frac{a(1+\epsilon)}{a(1-\epsilon)} = \frac{1+0.055}{1-0.055} = 1.12.$$

6.33

a)   The ratio of the lengths of the major and minor axes is

$$\frac{2a}{2b} = \frac{a}{b}.$$

In an ellipse, a, b, and c (the distance from center to either focus) are related by $a^2 = b^2 + c^2$, as can be seen from the following diagram:

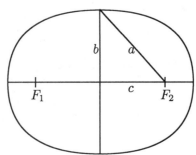

Therefore,

$$b^2 = a^2 - c^2 = a^2 - (\epsilon a)^2 = a^2(1 - \epsilon^2) \implies b = a\sqrt{1 - \epsilon^2}.$$

Hence the ratio of the lengths of major to minor axes is

$$\frac{a}{b} = \frac{1}{\sqrt{1 - \epsilon^2}}.$$

b)   First, use the result from part a), to solve for ϵ in terms of $\dfrac{a}{b}$.

$$\frac{a}{b} = \frac{1}{\sqrt{1-\epsilon^2}} \implies \sqrt{1-\epsilon^2} = \frac{b}{a} \implies 1 - \epsilon^2 = \frac{b^2}{a^2} \implies \epsilon^2 = 1 - \frac{b^2}{a^2} \implies \epsilon = \sqrt{1 - \frac{b^2}{a^2}} = \sqrt{1 - \frac{1}{\left(\frac{a}{b}\right)^2}}.$$

If the ratio is 1.01, then the last formula yields

$$\epsilon = \sqrt{1 - \frac{1}{(1.01)^2}} = 0.140.$$

If the ratio is 1.10, then the same formula yields $\epsilon = 0.417$.

6.37 Example 6:7 from the text showed that the ratio of the speeds is

$$\frac{v_{\text{perihelion}}}{v_{\text{aphelion}}} = \frac{1+\epsilon}{1-\epsilon} = \frac{1+0.0167}{1-0.0167} = \frac{1.0167}{0.9833} = 1.034.$$

6.41 Use Kepler's third law with customized units:

$$T^2 = 1\frac{\text{y}^2}{\text{AU}^3}(a^3) = 1\frac{\text{y}^2}{\text{AU}^3}(2.8\text{ AU})^3 = 22\text{ y}^2 \implies T = 4.7\text{ y}.$$

6.45

a) Use Kepler's third law with customized units:

$$T^2 = 1\frac{\text{y}^2}{\text{AU}^3}(a^3) \implies (75.6\text{ y})^2 = 1\frac{\text{y}^2}{\text{AU}^3}(a^3) \implies a^3 = 5.72 \times 10^3\text{ AU}^3 \implies a = 17.9\text{ AU}.$$

b) The perihelion distance is given by

$$r_{\text{perihelion}} = a(1-\epsilon) \implies 0.57\text{ AU} = (17.9\text{ AU})(1-\epsilon) \implies \epsilon = 1 - \frac{0.57\text{ AU}}{17.9\text{ AU}} = 0.968.$$

c) The aphelion distance is

$$r_{\text{aphelion}} = a(1+\epsilon) = (17.9\text{ AU})(1+0.968) = 35.3\text{ AU}.$$

d) The maximum speed occurs at the perihelion point, while the minimum speed occurs at the aphelion point. In Example 6:7 the ratio of these speeds was found to be

$$\frac{v_{\text{perihelion}}}{v_{\text{aphelion}}} = \frac{1+\epsilon}{1-\epsilon} = \frac{1+0.968}{1-0.968} \approx 62,$$

which is 6×10^1 if we keep only one significant figure.

6.49

a) Use the geometry shown below.

The location of the center of mass is found from

$$mr_1 = Mr_2,$$

where m and M are the masses of the Moon and Earth respectively and $r_1 + r_2 = r$, the distance between the centers of the Moon and the Earth. Substitute $r - r_2$ for r_1, $80m$ for M, and then solve for r_2.

$$m(r - r_2) = 80mr_2 \implies r_2 = \left(\frac{1}{1+80}\right)(3.84 \times 10^5\text{ km}) = 4.7 \times 10^3\text{ km}.$$

This distance places the center of mass of the Earth-Moon system inside the Earth.

b) Since the radius of the Earth is 6.37×10^3 km, the center of mass of the Earth-Moon system is

$$6.37 \times 10^3\text{ km} - 4.7 \times 10^3\text{ km} = 1.7 \times 10^3\text{ km}$$

below the surface of the Earth along the Earth-Moon line.

6.53

a)   The magnitude of the gravitational force is

$$F_1 = \frac{G\left(\dfrac{M}{4}\right)\left(\dfrac{M}{4}\right)}{(2r)^2} = \frac{GM^2}{64r^2}.$$

b)   The magnitude of the gravitational force is

$$F_2 = \frac{GM\left(\dfrac{M}{4}\right)}{r^2} = \frac{GM^2}{4r^2}.$$

c)   The total gravitational force on Ψ is the vector sum of the two forces. Since the forces act along the same line and point in the same direction, the magnitude of the total gravitational force is the simple sum of the magnitudes of the individual forces:

$$F_{\text{total}} = F_1 + F_2 = \frac{GM^2}{64r^2} + \frac{GM^2}{4r^2} = \left(\frac{17}{64}\right)\left(\frac{GM^2}{r^2}\right).$$

d)   The star completes one circumference of its orbit during the period T. Hence the speed of the star is

$$v = \frac{2\pi r}{T}.$$

e)   Apply Newton's second law to the star Ψ. Since the star is in a circular orbit, the acceleration is the centripetal acceleration. Use the magnitudes of the total force and acceleration:

$$F_{\text{total}} = ma \implies \left(\frac{17}{64}\right)\left(\frac{GM^2}{r^2}\right) = \left(\frac{M}{4}\right)\left(\frac{v^2}{r}\right) \implies \left(\frac{17}{16}\right)\left(\frac{GM}{r}\right) = v^2.$$

Now substitute $\dfrac{2\pi r}{T}$ for v:

$$\left(\frac{17}{16}\right)\left(\frac{GM}{r}\right) = \left(\frac{2\pi r}{T}\right)^2 \implies T^2 = \frac{64\pi^2 r^3}{17GM}.$$

6.57   Each star traverses the circumference of its orbit in one period. Hence, the speed of each star is

$$v = \frac{2\pi r}{T} \implies r = \frac{vT}{2\pi} = \frac{(250 \times 10^3 \text{ m/s})(14.4 \text{ d})}{2\pi}\left(\frac{8.64 \times 10^4 \text{ s}}{d}\right) = 4.95 \times 10^{10} \text{ m}.$$

The separation of the two stars is twice the radius of their orbits. The force on either star is the gravitational force of the other star. Apply Newton's second law to either star:

$$F_{\text{total}} = ma \implies \frac{Gmm}{(2r)^2} = m\frac{v^2}{r} \implies \frac{Gm}{4r} = v^2.$$

Now solve for m:

$$m = \frac{4rv^2}{G} = \frac{4(4.95 \times 10^{10} \text{ m})(2.50 \times 10^5 \text{ m/s})^2}{6.67 \times 10^{-11} \text{ N} \cdot \text{m}^2/\text{kg}^2} = 1.86 \times 10^{32} \text{ kg}.$$

The solar mass is 1.99×10^{30} kg. Hence the mass of each of these stars is

$$m = (1.86 \times 10^{32} \text{ kg})\left(\frac{\text{solar mass}}{1.99 \times 10^{30} \text{ kg}}\right) = 93.5 \text{ solar masses}.$$

6.61

a)   First find the number of meters in a light year(LY).

$$1 \text{ LY} = (3.00 \times 10^8 \text{ m/s})(365.25 \text{ d}) \left(\frac{8.6400 \times 10^4 \text{ s}}{\text{d}} \right) = 9.47 \times 10^{15} \text{ m}.$$

Now convert the radius r of the orbit of the solar system about the center of the galaxy from light years to meters:

$$r = (2.5 \times 10^4 \text{ LY}) \left(\frac{9.47 \times 10^{15} \text{ m}}{\text{LY}} \right) = 2.4 \times 10^{20} \text{ m}.$$

The time required to complete one orbit is the circumference of the orbit divided by the speed,

$$T = \frac{2\pi r}{v} = \frac{(2\pi)(2.4 \times 10^{20} \text{ m})}{230 \times 10^3 \text{ m/s}} = 6.6 \times 10^{15} \text{ s} = (6.6 \times 10^{15} \text{ s}) \left(\frac{y}{3.16 \times 10^7 \text{ s}} \right) = 2.1 \times 10^8 \text{ y}.$$

b)   The number of times the solar system has orbited the center of the galaxy is the age of the solar system divided by the period for one orbit:

$$\frac{5.0 \times 10^9 \text{ y}}{2.1 \times 10^8 \text{ y}} = 24.$$

c)   Apply Newton's second law to the solar system (of mass m, which is essentially the mass of the Sun since the Sun contains 99.99% of all the mass in the solar system). Let M be the mass inside our galactic orbit. The total force on the solar system is the gravitational force of M on m. Since the orbit of the solar system is assumed to be circular, the acceleration is a centripetal acceleration. Thus

$$F_{\text{total}} = ma \implies \frac{GMm}{r^2} = m\frac{v^2}{r} \implies M = \frac{rv^2}{G} = \frac{(2.4 \times 10^{20} \text{ m})(230 \times 10^3 \text{ m/s})^2}{6.67 \times 10^{-11} \text{ N} \cdot \text{m}^2/\text{kg}^2} = 1.9 \times 10^{41} \text{ kg}.$$

Convert this mass to units of solar masses by dividing by the mass of the Sun:

$$M = 1.9 \times 10^{41} \text{ kg} = (1.9 \times 10^{41} \text{ kg}) \left(\frac{\text{solar mass}}{1.99 \times 10^{30} \text{ kg}} \right) = 9.5 \times 10^{10} \text{ solar masses}.$$

This is roughly 100 billion solar masses!

d)   If you assume the material located beyond the galactic orbit of the solar system is in a spherically symmetric distribution about the center of the galaxy (an unrealistic assumption, in fact), then this more distant material has no effect on the orbit of the solar system because the gravitational force of such a spherical shell on a mass within it is zero.

6.65

a)   Use the geometry shown below, where d is the distance between the centers of Earth and Moon, and r is the distance from the center of the Earth to the L_1 Lagrangian point.

At the L_1 Lagrangian point, the gravitational field of the Earth and that of the Moon have equal magnitudes, but point in opposite directions and so vector sum to zero. Equate the magnitudes of the two fields:

$$\frac{GM_{\text{Earth}}}{r^2} = \frac{GM_{\text{Moon}}}{(d-r)^2} \implies \frac{M_{\text{Earth}}}{M_{\text{Moon}}} = \frac{r^2}{(d-r)^2} \implies 81.3 = \frac{r^2}{(d-r)^2} \implies 9.02 = \frac{r}{d-r} \implies r = 0.900d.$$

Thus L_1 is located 9/10 of the way from the Earth to the Moon.

b)   If a mass is moved slightly toward either the Moon or the Earth, the mass will not return to the L_1 point because one or the other of the gravitational forces on the mass will be of greater magnitude than the other, so the total force on the mass will be nonzero. The nonzero total force will accelerate the mass toward either the Moon or the Earth. Thus, although L_1 is an *equilibrium point*, it is an *unstable* equilibrium point.

6.69   The magnitude of the gravitational field of the Earth of mass M at a distance r from its center is

$$g(r) = -\frac{GM}{r^2}.$$

To find the variation of $g(r)$ with r, take its derivative with respect to r:

$$\frac{d}{dr}g(r) = 2\frac{GM}{r^3}.$$

So, for small changes Δr in r, the resulting change Δg in $g(r)$ is given approximately by

$$\Delta g = g(r + \Delta r) - g(r) \approx 2\frac{GM}{r^3}\Delta r$$

Therefore, solving for Δr,

$$\Delta r \approx \frac{r^3}{2GM}\Delta g = \frac{(6.37 \times 10^6 \text{ m})^3(9.832 \text{ m/s}^2 - 9.780 \text{ m/s}^2)}{2(6.67 \times 10^{-11} \text{ N} \cdot \text{m}^2/\text{kg}^2)(5.98 \times 10^{24} \text{ kg})} = 1.7 \times 10^4 \text{ m} = 17 \text{ km}.$$

The difference between the polar and equatorial diameters is twice the difference in the polar and equatorial radii, or 34 km This result is smaller than the actual difference between the equatorial and polar diameters, which is about 44 km because we have used a spherical model for the Earth. The calculation does provide a good order of magnitude estimate of the difference.

6.73   The flux of the total gravitational field through a closed surface enclosing the Moon is $-4\pi G$ times the mass of the Moon:

$$-4\pi GM_{\text{Moon}} = -4\pi(6.67 \times 10^{-11} \text{ N} \cdot \text{m}^2/\text{kg}^2)(7.36 \times 10^{22} \text{ kg}) = -6.17 \times 10^{13} \text{ m}^3/\text{s}^2.$$

If another surface encloses both the Moon and the Earth, the flux of the gravitational field through this surface is $-4\pi G(M_{\text{Moon}} + M_{\text{Earth}})$ The increase in the flux is therefore

$$-4\pi GM_{\text{Earth}} = -4\pi(6.67 \times 10^{-11} \text{ N} \cdot \text{m}^2/\text{kg}^2)(5.98 \times 10^{24} \text{ kg}) = -5.01 \times 10^{15} \text{ m}^3/\text{s}^2.$$

Chapter 7

Hooke's Force Law and Simple Harmonic Oscillation

7.1   The magnitude of the force you exert on a spring is proportional to its extension:

$$F = kx \implies 5.00 \text{ N} = k(0.10 \text{ m}) \implies k = \frac{5.00 \text{ N}}{0.10 \text{ m}} = 50 \text{ N/m}.$$

7.5   Think of the spring as being located to the left of the mass and choose a coordinate system with $\hat{\mathbf{i}}$ in the direction the spring stretches, $\hat{\mathbf{j}}$ pointing up, and origin where the mass was located before the spring was stretched. So x is the distance that the spring has been stretched.

The forces acting on the mass when it is on the verge of slipping are:

1. the weight $\vec{\mathbf{w}}$ of the mass; $\vec{\mathbf{w}} = (-mg)\hat{\mathbf{j}}$;

2. the normal force $\vec{\mathbf{N}}$ of the surface, which is along $\hat{\mathbf{j}}$;

3. the force of the spring on the mass, equal to $(-kx)\hat{\mathbf{i}}$;

4. the maximum value of the force of static friction $\vec{\mathbf{f}}_{s \text{ max}}$. It is directed to oppose slippage, and so is along $\hat{\mathbf{i}}$. Its magnitude is $\mu_s N$.

The total force on the system is zero when it is on the verge of slipping, so

x direction	y direction
$F_{x \text{ total}} = ma_x \implies \mu_s N - kx = m(0 \text{ m/s}^2)$	$F_{y \text{ total}} = ma_y \implies N - mg = m(0 \text{ m/s}^2)$
$\implies \mu_s N = kx$	$\implies N = mg.$

Take the expression for N from the y-direction and substitute it into the last x-direction equation:

$$\mu_s mg = kx \implies k = \frac{\mu_s mg}{x} = \frac{0.25(15.5 \text{ kg})(9.81 \text{ m/s}^2)}{0.150 \text{ m}} = 2.5 \times 10^2 \text{ N/m}.$$

7.9

a)   Choose $\hat{\mathbf{i}}$ pointing to the right. Let A be the point at which the two springs are joined. Stretch and hold the spring system in the $\hat{\mathbf{i}}$ direction by moving the mass m to the right. Let x' be the distance that point A moves, and let x be the distance that the mass m moves. Then spring 1 exerts a force on point A equal to $-k_1 x'\hat{\mathbf{i}}$, while spring 2 exerts a force equal to $k_2(x - x')\hat{\mathbf{i}}$. Since point A is at rest, the vector sum of these two forces must be zero,

$$-k_1 x'\hat{\mathbf{i}} + k_2(x - x')\hat{\mathbf{i}} = \mathbf{0} \text{ N} \implies -k_1 x' + k_2(x - x') = 0 \text{ N} \implies x' = \frac{k_2}{k_1 + k_2}x.$$

b)   Let $\vec{\mathbf{F}}$ be the force you are exerting to hold the mass in position. The forces acting on the mass m are

1. the force $\vec{\mathbf{F}}$ that you exert;

2. the force of spring 2 on m, which equals $-k_2(x - x')\hat{\mathbf{i}}$, since $(x - x')$ is the extension of the spring and the force is acting in the $-\hat{\mathbf{i}}$ direction.

Since the mass is not accelerating, the total force on it is zero. Therefore, using part a)

$$\vec{\mathbf{F}} - k_2(x - x')\hat{\mathbf{i}} = \mathbf{0} \text{ N} \implies \vec{\mathbf{F}} - k_2 x \left(1 - \frac{k_2}{k_1 + k_2} \right) \hat{\mathbf{i}} = \mathbf{0} \text{ N} \implies \vec{\mathbf{F}} - \frac{k_1 k_2}{k_1 + k_2} x \hat{\mathbf{i}} = \mathbf{0} \text{ N}.$$

If you replace the dual springs with a single spring, with force constant k_e, that is stretched and held at a distance x, then

$$\vec{\mathbf{F}} - k_e x \hat{\mathbf{i}} = \mathbf{0} \text{ N}.$$

Compare this with the previous equation and find

$$k_e = \frac{k_1 k_2}{k_1 + k_2}.$$

This result is sometimes stated as "spring constants in series have reciprocals that add," since

$$k_e = \frac{k_1 k_2}{k_1 + k_2} \iff \frac{1}{k_e} = \frac{1}{k_1} + \frac{1}{k_2}.$$

7.13

a)   The position of the oscillator as a function of time is $x(t) = A\cos(\omega t + \theta)$, and the velocity component is $v_x(t) = -A\omega \sin(\omega t + \theta)$. The speed is the absolute value of the velocity component. Since the maximum value of $|\sin(\omega + \theta)|$ is 1, the maximum speed is

$$v_{max} = A\omega \implies \omega = \frac{v_{max}}{A} = \frac{2.00 \text{ m/s}}{0.0500 \text{ m}} = 40.0 \text{ rad/s}.$$

b)   The frequency is

$$\nu = \frac{\omega}{2\pi} = \frac{40.0 \text{ rad/s}}{2\pi} = 6.37 \text{ Hz}.$$

c)   The general expression for the position is $x(t) = A\cos(\omega t + \theta)$. When $t = 0$ s, $x = 0$ m, so

$$0 \text{ m} = A\cos[\omega(0 \text{ s}) + \theta] = A\cos\theta.$$

The amplitude A is not zero, so we must have $\cos\theta = 0$, and therefore $\theta = \dfrac{\pi}{2}$ rad. Hence

$$x(t) = A\cos(\omega t + \theta) = (0.0500 \text{ m})\cos[(40.0 \text{ rad/s})t + \frac{\pi}{2} \text{ rad}].$$

d)   When $t = 1.00$ s, the position of the mass is

$$x = (0.0500 \text{ m})\cos[(40.0 \text{ rad/s})(1.00 \text{ s}) + \frac{\pi}{2} \text{ rad}] = (0.0500 \text{ m})\cos(41.6 \text{ rad}) = -0.0363 \text{ m}.$$

7.17

a) The angular frequency is found from

$$\omega = \sqrt{\frac{k}{m}} = \sqrt{\frac{200 \text{ N/m}}{1.50 \text{ kg}}} = 11.5 \text{ rad/s}.$$

When $t = 0$ s, the mass is released at $x = 0.100$ m, so the general equation for x becomes

(1) $$0.100 \text{ m} = A\cos[\omega(0\text{ s}) + \phi] = A\cos\phi.$$

Likewise, when $t = 0$ s, the velocity component is $v_x = 2.00$ m/s. The velocity at any instant is

$$v_x(t) = \frac{d}{dt}x(t) = -A\omega\sin(\omega t + \phi).$$

So

(2) $$2.00 \text{ m/s} = -A\omega\sin[\omega(0\text{ s}) + \phi] = -A(11.5 \text{ rad/s})\sin\phi.$$

Divide equation (2) by equation (1).

$$\frac{2.00 \text{ m/s}}{0.100 \text{ m}} = \frac{-A(11.5 \text{ rad/s})\sin\phi}{A\cos\phi} \implies 20.0 \text{ s}^{-1} = -11.5 \text{ rad/s} \tan\phi$$

$$\implies \tan\phi = -1.74$$

$$\implies \phi = -1.05 \text{ rad}.$$

Now use this value for ϕ in equation (1) to find A:

$$0.100 \text{ m} = A\cos\phi = A\cos(-1.05 \text{ rad}) \implies A = 0.201 \text{ m}.$$

Hence

$$x(t) = (0.201 \text{ m})\cos[(11.5 \text{ rad/s})t - 1.05 \text{ rad}].$$

b) The period T of the oscillation is the inverse of the frequency ν:

$$T = \frac{1}{\nu} = \frac{2\pi}{\omega} = \frac{2\pi}{11.5 \text{ rad/s}} = 0.546 \text{ s}.$$

c) The velocity component at any time is

$$v_x(t) = \frac{d}{dt}x(t) = -A\omega\sin(\omega t + \phi).$$

Since the maximum magnitude of the sine is 1, the maximum speed is

$$v_{max} = \omega A = (11.5 \text{ rad/s})(0.201 \text{ m}) = 2.31 \text{ m/s}.$$

The acceleration component at any time is

$$a_x(t) = \frac{d}{dt}v(t) = -A\omega^2\cos(\omega t + \phi).$$

Since the maximum magnitude of the cosine is 1, the maximum magnitude of the acceleration is

$$a_{max} = \omega^2 A = (11.5 \text{ rad/s})^2(0.201 \text{ m}) = 26.6 \text{ m/s}^2.$$

d) A plot of $x(t)$ versus t for two periods is shown below.

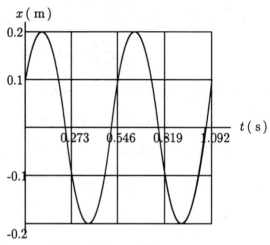

7.21 In Example 7:8, the position coordinate of the oscillator was

$$x(t) = (0.0707 \text{ m}) \cos[(7.07 \text{ rad/s})t - \frac{\pi}{2} \text{ rad}].$$

The velocity component is

$$v_x(t) = \frac{d}{dt}x(t) = -(0.5 \text{ m/s}) \sin[(7.07 \text{ rad/s})t - \frac{\pi}{2} \text{ rad}].$$

The acceleration component is

$$a_x(t) = \frac{d}{dt}v(t) = -(3.54 \text{ m/s}^2) \cos[(7.07 \text{ rad/s})t - \frac{\pi}{2} \text{ rad}].$$

Graphs of $v_x(t)$ and $a_x(t)$ appear below.

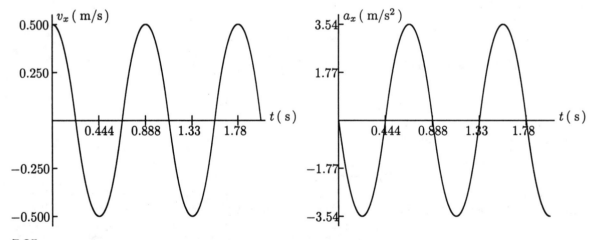

7.25

a) The period of the x_1 oscillation is

$$T_1 = \frac{2\pi}{\omega_1} = \frac{2\pi}{0.250 \text{ rad/s}} = 25.1 \text{ s}.$$

The period of the x_2 oscillation is

$$T_2 = \frac{2\pi}{\omega_2} = \frac{2\pi}{0.350 \text{ rad/s}} = 18.0 \text{ s}.$$

b)   Graphs of $x_1(t)$ and $x_2(t)$ are shown below. The places where the oscillators have the same position are the places where the graphs intersect. The first six such locations are estimated from the graph to occur when t is approximately 1 s, 9 s, 19 s, 29 s, 39 s, and 49 s. The interval between these instants is about 10 s.

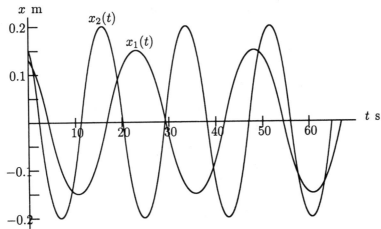

c)   Change the amplitude of the x_1 oscillation to 0.200 m and replot both oscillations, as shown below. The times of intersection change as well as the intervals between them.

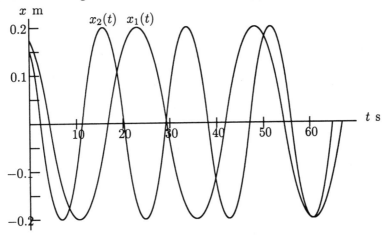

7.29   The forces on the mass are:

    1. its weight $\vec{\mathbf{w}}$, with magnitude mg, directed downward;

    2. the force of the spring, directed upward.

Choose a coordinate system with $\hat{\mathbf{i}}$ pointing down, in the same direction as $\vec{\mathbf{w}}$, and with origin where the object would be if it's mass were zero. Let x be the amount by which the spring is stretched. Then the total force on the object is $mg\hat{\mathbf{i}} + kx\hat{\mathbf{i}}$, where k is the spring constant. Since the acceleration of the mass is zero in its equilibrium position,

$$\vec{\mathbf{F}}_{\text{total}} = m\vec{\mathbf{a}} \implies mg\hat{\mathbf{i}} - kx\hat{\mathbf{i}} = \mathbf{0}\text{ N} \implies mg - kx = 0\text{ N} \implies x = \frac{mg}{k} = \frac{(2.00\text{ kg})(9.81\text{ m/s}^2)}{100\text{ N/m}} = 1.96\text{ m}.$$

The distance x is measured from the original unstretched length of the spring (without the mass on it). Hence the new equilibrium position of the mass, about which it will execute simple harmonic oscillations, is

$$0.196\text{ m} + 0.300\text{ m} = 0.496\text{ m}$$

from the fixed end of the spring.

7.33

a) The angular frequency of the oscillation is

$$\omega = \sqrt{\frac{k}{m}} = \sqrt{\frac{50.0 \text{ N/m}}{0.750 \text{ kg}}} = 8.16 \text{ rad/s}.$$

So the frequency is

$$\nu = \frac{\omega}{2\pi} = \frac{8.16 \text{ rad/s}}{2\pi} = 1.30 \text{ Hz}.$$

b) The tension in the string is a minimum at the highest point of the oscillation.

c) The tension in the string is a maximum at the lowest point of the oscillation.

d) The acceleration component of the oscillator at any time is

$$a_x(t) = \frac{d^2}{dt^2}x(t) = -A\omega^2 \cos(\omega t + \phi).$$

Since the maximum magnitude of the cosine is 1, the maximum magnitude of the acceleration is $a_{max} = A\omega^2$. This acceleration occurs when the oscillator has its maximum displacement. Take the 0.250 kg mass as the system. The forces on this mass are:

1. its weight $\vec{w}$, of magnitude m g, directed downward; and

2. the force $\vec{T}$ of the string on m, directed upward.

Choose a coordinate system with $\hat{i}$ directed downward, parallel to $\vec{w}$. At the highest point of the oscillation

$$F_{x \text{ total}} = ma_x \implies mg - T = ma_{max}.$$

Note that T is zero if $a_{max} = g$. Hence

$$A\omega^2 = g \implies A = \frac{g}{\omega^2} = \frac{9.81 \text{ m/s}^2}{(8.16 \text{ rad/s})^2} = 0.147 \text{ m}.$$

7.37

a) The velocity component at any time is

$$v_x(t) = \frac{d}{dt}x(t) = -A\omega \sin(\omega t + \phi).$$

Since the maximum magnitude of the sine is 1, the maximum speed is $v_{max} = A\omega$.
    The angular frequency of the oscillation is

$$\omega = \sqrt{\frac{k}{m}} = \sqrt{\frac{350 \text{ N/m}}{1.5 \text{ kg}}} = 15 \text{ rad/s}.$$

The amplitude of the oscillation is given as $A = 0.10 \text{ m}$. Therefore,

$$v_{max} = (0.10 \text{ m})(15 \text{ rad/s}) = 1.5 \text{ m/s}.$$

b) The acceleration component of the oscillator at any time is

$$a_x(t) = \frac{d^2}{dt^2}x(t) = -A\omega^2 \cos(\omega t + \phi).$$

Since the maximum magnitude of the cosine is 1, the maximum magnitude of the acceleration is

$$a_{max} = A\omega^2 = (0.10 \text{ m})(15 \text{ rad/s})^2 = 23 \text{ m/s}^2.$$

7.41 Let $\theta'(t)$ be the angle that the particle makes with the x' axis at time t. Then, from the Figure P.41 in the text, the projection of the particle onto the x' axis is

(1) $$x'(t) = R \cos \theta'(t).$$

At any time t, the angle the particle makes with the x'-axis is 30° less than the angle it makes with the x axis. But at time t the angle it makes with the x-axis is just ωt. Hence,

(2) $$\theta'(t) = \omega t - 30° = \omega t - \frac{\pi}{6} \text{ rad}.$$

Combining equations (1) and (2) we have

$$x'(t) = R \cos \left(\omega t - \frac{\pi}{6} \text{ rad} \right).$$

7.45

a)   The period of the dog is the time for one round trip, which is 10 s.

b)   The frequency ν is the inverse of the period:

$$\nu = \frac{1}{T} = \frac{1}{10 \text{ s}} = 0.10 \text{ Hz}.$$

c)   It is unlikely that the motion of the dog is simple harmonic motion. Dogs run at essentially constant speed, except when stopping and reversing direction.

7.49   Set up the geometry as shown in the sketch below:

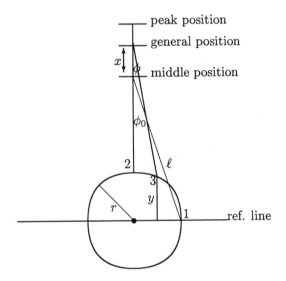

The point where the tie-rod is connected to the rim of the shaft is in circular motion. When the connection point is at position 1, the piston is at the midpoint of its oscillation, located a distance $\ell \cos \phi_0$ above a reference horizontal line through the center of the circle. When the connection point at position 2, the piston is at the peak of its upward motion, a distance $\ell + r$ above the reference line. When the connection point at a general position, such as 3 in the sketch, a distance y above the reference line, the piston is a distance x above the midpoint of its oscillation. From the geometry, the following expressions are equivalent representations of the same distance:

(1) $$x + \ell \cos \phi_0 = y + \ell \cos \phi$$

Note that $\ell \cos \phi_0$ does not vary with time, but x, y, and ϕ do vary with time. Differentiate equation (1) with respect to t:

$$\frac{dx}{dt} = -\ell \sin \phi \left(\frac{d\phi}{dt} \right) + \left(\frac{dy}{dt} \right).$$

Now take the second derivative:

(2)
$$\frac{d^2}{dt^2}x = -\ell \sin \phi \left(\frac{d^2}{dt^2}\phi \right) - \ell \cos \phi \left(\frac{d}{dt}\phi \right)^2 + \frac{d^2}{dt^2}y.$$

Since y is the projection of the motion of the point in uniform circular motion on the vertical line, y undergoes simple harmonic oscillation with an angular frequency ω equal to the angular speed of the uniform circular motion. In other words, y obeys the simple harmonic oscillator equation:

$$\frac{d^2}{dt^2}y + \omega^2 y = 0 \implies \frac{d^2}{dt^2}y = -\omega^2 y.$$

Make this substitution for the second derivative of y in equation (2) above:

$$\frac{d^2}{dt^2}x = -\ell \sin \phi \left(\frac{d^2}{dt^2}\phi \right) - \ell \cos \phi \left(\frac{d\phi}{dt} \right)^2 - \omega^2 y.$$

Now use equation (1) to substitute for y:

$$\frac{d^2}{dt^2}x = -\ell \sin \phi \left(\frac{d^2}{dt^2}\phi \right) - \ell \cos \phi \left(\frac{d\phi}{dt} \right)^2 - \omega^2 (x + \ell \cos \phi_0 - \ell \cos \phi).$$

Rearrange this slightly:

$$\frac{d^2}{dt^2}x + \omega^2 x = -\ell \sin \phi \left(\frac{d^2}{dt^2}\phi \right) - \ell \cos \phi \left(\frac{d\phi}{dt} \right)^2 - \omega^2 (\ell \cos \phi_0 - \ell \cos \phi).$$

The vertical coordinate x of the piston does not satisfy the simple harmonic oscillator differential equation because

$$\frac{d^2}{dt^2}x + \omega^2 x \neq 0.$$

Therefore, the motion of the piston is not simple harmonic oscillation.

7.53   The period T of a simple pendulum of length ℓ is given by

$$T = 2\pi \sqrt{\frac{\ell}{g}} = 2\pi \sqrt{\frac{67.5 \text{ m}}{9.81 \text{ m/s}^2}} = 16.5 \text{ s}.$$

7.57   Let T be the length of the interval we seek. The longer pendulum has the greater period of oscillation. The small oscillations of a simple pendulum are simple harmonic oscillations. During each period of each oscillation, the argument of the cosine function describing each oscillation increases by 2π rad. Hence, for each pendulum, the number of radians per second each completes is $\frac{2\pi}{T_1}$ and $\frac{2\pi}{T_2}$ respectively.

Since $T_1 > T_2$, the second pendulum completes more radians per second than the first. The difference $\frac{2\pi}{T_2} - \frac{2\pi}{T_1}$ between the two quantities is the number of radians per second that the second pendulum is ahead of the first in its oscillation. When the two pendulums are "back in step," the second pendulum has gained 2π rad on the first. Hence the above difference must be $\frac{2\pi \text{ rad}}{T}$. That is,

$$\frac{2\pi}{T_2} - \frac{2\pi}{T_1} = \frac{2\pi}{T} \implies \frac{1}{T_2} - \frac{1}{T_1} = \frac{1}{T} \implies T = \frac{T_1 T_2}{T_1 - T_2}.$$

7.61 Since the clock runs slow, the period of its pendulum is longer at the new location, and since ℓ is unchanged we anticipate a smaller value for g. The period T of a simple pendulum of length ℓ is $T = 2\pi\sqrt{\dfrac{\ell}{g}}$. To see how the period is affected by changes in g, take the derivative of T with respect to g:

$$\frac{d}{dg}T = 2\pi\ell^{1/2}\left(-\frac{1}{2}g^{-3/2}\right) = -\left(2\pi\sqrt{\frac{\ell}{g}}\right)\frac{1}{2g} = -\frac{T}{2g}.$$

So, for a small change Δg in g and the resulting change ΔT in T we have approximately

$$\Delta T \approx -\frac{T}{2g}\Delta g \implies \Delta g \approx -2g\frac{\Delta T}{T}.$$

The fractional increase in the period is 89 s divided by the number of seconds per day:

$$\frac{\Delta T}{T} = \frac{89\text{ s}}{8.64 \times 10^4\text{ s}} = 1.0 \times 10^{-3} \implies \Delta g \approx -2g\frac{\Delta T}{T} = -2(9.81\text{ m/s}^2)(1.0 \times 10^{-3}) = -2.0 \times 10^{-2}\text{ m/s}^2.$$

The magnitude of the acceleration due to gravity at the new location is therefore

$$g_{\text{new}} = g + \Delta g \approx 9.81\text{ m/s}^2 - 0.020\text{ m/s}^2 = 9.79\text{ m/s}^2.$$

7.65 If the one-way trip is 1.00 min, the period is 2.00 min. In Section 7.7 of the text, it was shown that the period T of a mass oscillating through a uniform sphere of mass M and radius R is

$$T = 2\pi\sqrt{\frac{R^3}{GM}}.$$

So

$$R = \sqrt[3]{\frac{T^2GM}{4\pi^2}} = \sqrt[3]{\frac{(120\text{ s})^2(6.67 \times 10^{-11}\text{ N}\cdot\text{m}^2/\text{kg}^2)(5.98 \times 10^{24}\text{ kg})}{4\pi^2}} = 5.26 \times 10^5\text{ m} = 526\text{ km}.$$

Recall that the actual radius of the Earth is 6 370 km. Hence, in order for a mass to make the one way trip through the hypothetical tunnel in a minute, the Earth would have to have the same mass but a radius equal to only 8.3% of its actual radius.

7.69 Once you have determined the time for the amplitude to decrease to half its initial value, use the result of problem 68 to determine β. From problem 68 we have

$$t = \frac{2m}{\beta}\ln 2,$$

so

$$\beta = \frac{2m}{t}\ln 2.$$

7.73 First convert the speed from km/h to m/s:

$$25\text{ km/h} = (25\text{ km/h})\left(\frac{10^3\text{ m}}{\text{km}}\right)\left(\frac{\text{h}}{3600\text{ s}}\right) = 6.9\text{ m/s}.$$

The bumps are spaced 0.50 m apart, so the car encounters

$$\frac{6.9\text{ m/s}}{0.50\text{ m/bump}} = 14\text{ bump/s}.$$

Since resonance occurs at this frequency, the natural oscillation frequency of the suspension system is 14 Hz.

Chapter 8

Work, Energy, and the CWE Theorem

8.1 The force is a constant force. Hence the work in each case can be found from the expression for the work done by a constant force:

$$W = \vec{\mathbf{F}} \bullet \Delta \vec{\mathbf{r}}.$$

a) The change in the position vector is $\Delta \vec{\mathbf{r}} = (5.00 \text{ m})\hat{\mathbf{i}}$, so

$$W = \vec{\mathbf{F}} \bullet \Delta \vec{\mathbf{r}} = (15.0 \text{ N})\hat{\mathbf{i}} \bullet (5.00 \text{ m})\hat{\mathbf{i}} = 75.0 \text{ J}.$$

b) The change in the position vector is $\Delta \vec{\mathbf{r}} = (2.00 \text{ m})\hat{\mathbf{j}}$, so

$$W = \vec{\mathbf{F}} \bullet \Delta \vec{\mathbf{r}} = (15.0 \text{ N})\hat{\mathbf{i}} \bullet (2.00 \text{ m})\hat{\mathbf{j}} = 0 \text{ J}.$$

c) The change in the position vector is

$$\Delta \vec{\mathbf{r}} = (4.33 \text{ m})\hat{\mathbf{i}} + (4.33 \text{ m})(\tan 30.0°)\hat{\mathbf{j}} = (4.33 \text{ m})\hat{\mathbf{i}} + (2.50 \text{ m})\hat{\mathbf{j}}.$$

Hence

$$W = \vec{\mathbf{F}} \bullet \Delta \vec{\mathbf{r}} = (15.0 \text{ N})\hat{\mathbf{i}} \bullet [(4.33 \text{ m})\hat{\mathbf{i}} + (2.50 \text{ m})\hat{\mathbf{j}}] = 65.0 \text{ J}.$$

d) The change in the position vector is

$$\Delta \vec{\mathbf{r}} = (-3.54 \text{ m})\hat{\mathbf{i}} + (3.54 \text{ m})(\tan 45°)\hat{\mathbf{j}} = (-3.54 \text{ m})\hat{\mathbf{i}} + (3.54 \text{ m})\hat{\mathbf{j}}.$$

Hence

$$W = \vec{\mathbf{F}} \bullet \Delta \vec{\mathbf{r}} = (15.0 \text{ N})\hat{\mathbf{i}} \bullet [(-3.54 \text{ m})\hat{\mathbf{i}} + (3.54 \text{ m})\hat{\mathbf{j}}] = -53.1 \text{ J}.$$

8.5

a) The force component varies linearly with x, so the graph is a straight line with a slope equal to the coefficient of x, which is -6.00 N/m. The "y" intercept is 0 N. Thus the graph should be downward sloping, with a slope of -6.00 N/m, and it should pass through the origin.

b) The x-component of the force is not constant, so you cannot use $W = \vec{\mathbf{F}} \bullet \Delta \vec{\mathbf{r}}$ to calculate its work. Revert to

$$W = \int \vec{\mathbf{F}} \bullet d\vec{\mathbf{r}} = \int_{9.00 \text{ m}}^{2.00 \text{ m}} \left((-6.00 \text{ N/m})\hat{\mathbf{i}} \right) \bullet (dz\hat{\mathbf{k}})$$

The unit vectors are perpendicular to each other, so the scalar product is zero; hence, $W = 0$ J.

71

8.9   The work done by a force component such as F_x is equal to the area under the graph of F_x versus x, between the initial and final positions. Here, the area is that of a semicircle. When evaluating the square of the radius of the circle, take one radius as the distance and the other as the force component.

$$W = \frac{\pi}{2}(30.0\text{ N})(5.0\text{ m}) = 2.4 \times 10^2\text{ J}.$$

8.13

a)   The acceleration is constant during the interval. The position of the system is

$$x(t) = x_0 + v_{x0}t + a_x\frac{t^2}{2} = x_0 + (3.00\text{ m/s}^2)\frac{t^2}{2}.$$

When $t = 4.00$ s

$$x = x_0 + (3.00\text{ m/s}^2)\frac{(4.00)^2}{2} = x_0 + 24.0\text{ m}.$$

So the change in the position vector of the system is

$$\Delta\vec{\mathbf{r}} = \vec{\mathbf{r}}_f - \vec{\mathbf{r}}_i = (x_0 + 24.0\text{ m})\hat{\mathbf{i}} - x_0\hat{\mathbf{i}} = 24.0\text{ m}\,\hat{\mathbf{i}}.$$

Since the acceleration component is constant, Newton's second law implies the total force component is constant, and

$$F_{x\text{ total}} = ma_x = (5.00\text{ kg})(3.00\text{ m/s}^2) = 15.0\text{ N}.$$

The work done by the constant force is

$$W = \vec{\mathbf{F}} \bullet \Delta\vec{\mathbf{r}} = (15.0\text{ N})\hat{\mathbf{i}} \bullet (24.0\text{ m})\hat{\mathbf{i}} = 360\text{ J}.$$

b)   The velocity component is constant, which implies the acceleration component is zero. Newton's second law then implies the total force component also is zero. Therefore, the work done by the total force is 0 J.

c)   Since the velocity component has a constant slope, the acceleration component is constant. To find the acceleration component, use the one-dimensional kinematics equation for constant acceleration,

$$v_x(t) = v_{x0} + a_xt = 5.00\text{ m/s} + a_xt.$$

When $t = 4.00$ s, the velocity component is zero, so

$$0\text{ m/s} = 5.00\text{ m/s} + a_x(4.00\text{ s}) \implies a_x = -1.25\text{ m/s}^2.$$

The position of the system at any time is

$$x(t) = x_0 + (5.00\text{ m/s})t - (1.25\text{ m/s}^2)\frac{t^2}{2}.$$

When $t = 4.00$ s, the x-coordinate is

$$x(t) = x_0 + (5.00\text{ m/s})(4.00\text{ s}) - (1.25\text{ m/s}^2)\frac{(4.00\text{ s})^2}{2} = x_0 + 10.00\text{ m}.$$

Thus the change in the position vector of the system is

$$\Delta\vec{\mathbf{r}} = \vec{\mathbf{r}}_f - \vec{\mathbf{r}}_i = (x_0 + 10.00\text{ m})\hat{\mathbf{i}} - x_0\hat{\mathbf{i}} = (10.00\text{ m})\hat{\mathbf{i}}.$$

Since the acceleration component is constant, Newton's second law implies the total force component is constant, so

$$F_{x\text{ total}} = ma_x = (5.00\text{ kg})(-1.25\text{ m/s}^2) = -6.25\text{ N}.$$

The work done by the force is

$$W = \vec{\mathbf{F}} \bullet \Delta\vec{\mathbf{r}} = (-6.25\text{ N})\hat{\mathbf{i}} \bullet (10.0\text{ m})\hat{\mathbf{i}} = -62.5\text{ J}.$$

d)   Since the velocity component has a constant slope, the acceleration component is constant. To find the acceleration component, use the one dimensional kinematics equation for constant acceleration,

$$v_x(t) = v_0 + a_x t = 0 \text{ m/s} + a_x t = a_x t.$$

When $t = 4.00$ s, the velocity component is 5.00 m/s, so

$$5.00 \text{ m/s} = a_x(4.00 \text{ s}) \implies a_x = 1.25 \text{ m/s}^2.$$

The position of the system at any time is

$$x(t) = x_0 + (0 \text{ m/s})t + (1.25 \text{ m/s}^2)\frac{t^2}{2}.$$

So, when $t = 4.00$ s,

$$x = x_0 + (0 \text{ m/s})(4.00 \text{ s}) + (1.25 \text{ m/s}^2)\frac{(4.00 \text{ s})^2}{2} = x_0 + 10.00 \text{ m}.$$

Therefore, the change in the position vector of the system is

$$\Delta \vec{\mathbf{r}} = \vec{\mathbf{r}}_f - \vec{\mathbf{r}}_i = (x_0 + 10.00 \text{ m})\hat{\mathbf{i}} - x_0\hat{\mathbf{i}} = (10.00 \text{ m})\hat{\mathbf{i}}.$$

Since the acceleration component is constant, Newton's second law implies the total force component is constant, and

$$F_{x \text{ total}} = ma_x = (5.00 \text{ kg})(1.25 \text{ m/s}^2) = 6.25 \text{ N}.$$

The work done by the constant force is

$$W = \vec{\mathbf{F}} \bullet \Delta \vec{\mathbf{r}} = (6.25 \text{ N})\hat{\mathbf{i}} \bullet (10.00 \text{ m})\hat{\mathbf{i}} = 62.5 \text{ J}.$$

8.17   The gravitational potential energy of a mass m located a distance r from another mass M is

$$\text{PE} = -\frac{GMm}{r}.$$

To find the change in the potential energy when r changes by a small amount, first find the derivative of the potential energy with respect to r.

$$\frac{d}{dr}\text{PE} = -GM\left(-\frac{1}{r^2}\right) = \frac{GMm}{r^2}.$$

Then, for small changes Δr in r, the corresponding change ΔPE in PE is

$$\Delta \text{PE} \approx \left(\frac{GMm}{r^2}\right)\Delta r.$$

8.21   The potential energy for Hooke's force law is

$$\text{PE} = \frac{kx^2}{2},$$

where x is the distance of the particle from its equilibrium position. Its derivative with respect to x is

$$\frac{d}{dx}\text{PE} = \frac{d}{dx}\left(\frac{kx^2}{2}\right) = kx.$$

So, for small changes Δx in x, the corresponding change in PE is

$$\Delta \text{PE} \approx (kx)\Delta x.$$

8.25   The Schwarzschild radius R_S of a black hole is

$$R_S = \frac{2GM}{c^2} = \frac{2(6.67 \times 10^{-11} \text{ N} \cdot \text{m}^2/\text{kg}^2)(75 \text{ kg})}{(3.00 \times 10^8 \text{ m/s})^2} = 1.1 \times 10^{-25} \text{ m}.$$

8.29   Apply the CWE theorem to the Dean, with the initial position taken as the point of release and the final position as the instant just before impact on the surface of the Earth. The gravitational force of the Earth is the only force acting on the Dean during the descent and its work is accounted for in the CWE theorem by the change in the appropriate gravitational potential energy. Since the position of the Dean changes over distances comparable to the radius of the Earth, you cannot use mgy as the gravitational potential energy function. Instead you must use the more general form

$$\text{PE} = -\frac{GMm}{r}.$$

Let R be the radius of the Earth, M the mass of the Earth, m the mass of the Dean, and v the speed of the Dean just before impact. There is zero work done by nonconservative forces (since there are none), so the CWE theorem becomes

$$0 \text{ J} = W_{\text{nonconservative}} = \Delta(\text{KE} + \text{PE}) = (\text{KE} + \text{PE})_\text{f} - (\text{KE} + \text{PE})_\text{i}$$

$$= \left[\frac{mv^2}{2} + \left(-\frac{GMm}{R}\right)\right] - \left[0 \text{ J} + \left(-\frac{GMm}{2R}\right)\right].$$

So $\dfrac{mv^2}{2} = \dfrac{GMm}{2R}$. Hence, solving for v,

$$v = \sqrt{\frac{GM}{R}} = \sqrt{\frac{(6.67 \times 10^{-11} \text{ N} \cdot \text{m}^2/\text{kg}^2)(5.98 \times 10^{24} \text{ kg})}{6.37 \times 10^6 \text{ m}}} = 7.91 \times 10^3 \text{ m/s} = 7.91 \text{ km/s}.$$

Converted to km/h, this is 28 500 km/h !

8.33

a)   The radius of the orbit is effectively the radius R of the Earth. The gravitational force of the Earth on the satellite is the only force acting on it (neglecting frictional effects). This force causes a centripetal acceleration. Using the magnitudes of the force and acceleration,

$$F_{\text{total}} = ma \implies \frac{GMm}{R^2} = m\frac{v^2}{R} \implies$$

$$v = \sqrt{\frac{GM}{R}} = \sqrt{\frac{(6.67 \times 10^{-11} \text{ N} \cdot \text{m}^2/\text{kg}^2)(5.98 \times 10^{24} \text{ kg})}{6.37 \times 10^6 \text{ m}}} = 7.91 \times 10^3 \text{ m/s} = 7.91 \text{ km/s}.$$

b)   From part a), the speed of the satellite is $v_{\text{grazing}} = \sqrt{\dfrac{GM}{R}}$. The escape speed from the Earth is $v_{\text{escape}} = \sqrt{\dfrac{2GM}{R}}$. Form the ratio

$$\frac{v_{\text{grazing}}}{v_{\text{escape}}} = \frac{\sqrt{\dfrac{GM}{R}}}{\sqrt{\dfrac{2GM}{R}}} = \frac{1}{\sqrt{2}} \approx 0.707.$$

8.37   The escape speed from the surface of a spherical mass M of radius R is

$$v_{\text{escape}} = \sqrt{\frac{2GM}{2R}}.$$

Squaring this gives

(1)
$$v_{\text{escape}}^2 = \frac{2GM}{2R}.$$

The mass of the asteroid is its average density times its volume.

$$M = \rho \frac{4\pi R^3}{3}.$$

Substitute this expression for M into equation (1),

$$v_{\text{escape}}^2 = \frac{2G\rho\dfrac{4\pi R^3}{3}}{R} = \frac{8\pi G\rho R^2}{3},$$

and solve for R:

$$R = \sqrt{\frac{3v_{\text{escape}}^2}{8\pi G\rho}} = \sqrt{\frac{3(10 \text{ m/s})^2}{8\pi(6.67 \times 10^{-11} \text{ N} \cdot \text{m}^2/\text{kg}^2)(5.3 \times 10^3 \text{ kg/m}^3)}} = 5.8 \times 10^3 \text{ m} = 5.8 \text{ km}.$$

8.41

a)   The potential energy of the Earth in the gravitational field of the Sun is

$$\text{PE} = \frac{-GMm}{r}.$$

b)   The only force acting on the Earth is the gravitational force of the Sun and the acceleration of the Earth is centripetal. Therefore

$$F_{\text{total}} = ma \implies \frac{GMm}{r^2} = \frac{mv^2}{r} \implies v = \sqrt{\frac{GM}{r}}.$$

c)   The kinetic energy of the Earth is

$$\text{KE} = \frac{mv^2}{2} = \frac{m\dfrac{GM}{r}}{2} = \frac{GMm}{2r}.$$

d)   The ratio is

$$\frac{\text{KE}}{\text{PE}} = \frac{\dfrac{GMm}{2r}}{\dfrac{-GMm}{r}} = -\frac{1}{2}.$$

This is a special case of a more general theorem in advanced mechanics called the *virial theorem* which relates the kinetic and potential energies of a body subject to a central force.

8.45

a)   Use the CWE theorem. During its descent, the forces acting on the vehicle are its weight $\vec{w}$ and the normal force $\vec{N}$. The normal force does zero work and the work done by the weight is accounted for by the potential energy term in the CWE theorem. The appropriate potential energy is mgy. Take $\hat{j}$ to point upward and choose the origin at point A. There are no nonconservative forces, so they do zero work. Take the initial position to be where the vehicle begins its descent at zero speed, and take point A as the final position. The CWE theorem becomes

$$0 \text{ J} = W_{\text{nonconservative}} = \Delta(\text{KE} + \text{PE}) = (\text{KE} + \text{PE})_{\text{f}} - (\text{KE} + \text{PE})_i = \left(\frac{mv_A^2}{2} + 0 \text{ J} \right) - (0 \text{ J} + mgy_{\text{i}})$$

$$\implies v_A = \sqrt{2gy_{\text{i}}} = \sqrt{2gh}.$$

b)   At point B, if the normal force of the track on the vehicle is zero, the only force on it is its weight, and this must provide the centripetal acceleration. Applying Newton's second law to the vehicle at point B, using the magnitudes of the vectors:

$$F_{\text{total}} = ma \implies mg = m\frac{v_B^2}{R} \implies v_B = \sqrt{gR}.$$

c)   Use the CWE theorem. Taking the initial position to be where the vehicle begins its journey, and the final position to be point B, we have

$$0 = W_{\text{nonconservative}} = \Delta(\text{KE} + \text{PE}) = (\text{KE} + \text{PE})_{\text{f}} - (\text{KE} + \text{PE})_i = \left(\frac{mv_B^2}{2} + mg(2R) \right) - (0 \text{ J} + mgh)$$

$$= \frac{mv_B^2}{2} + mg(2R) - mgh.$$

Substitute the expression for v_B from part b):

$$0 = \frac{mRg}{2} + mg(2R) - mgh \implies \frac{h}{R} = \frac{5}{2}.$$

d)   Since the total mechanical energy of the vehicle is conserved, the vehicle has the same speed when it exits the loop as it does when it enters the loop. Hence, from part a) the exit speed is $v_A = \sqrt{2gh}$.

e)   The work done by the gravitational force is the negative of the change in the gravitational potential energy of the vehicle,

$$W_{\text{gravity}} = -\Delta\text{PE} = -(\text{PE}_{\text{f}} - \text{PE}_{\text{i}}) = -(0 \text{ J} - mgh) = mgh.$$

f)   Only the answer to part e) depends upon the mass of the vehicle.

8.49   Choose a coordinate system with origin at the launch point, $\hat{i}$ in the direction of the horizontal component of the projectile's initial velocity, and $\hat{j}$ pointing up.

Apply the CWE theorem with the initial position at the place where the projectile is launched, and the final position at the highest point of its trajectory. At the highest point, the velocity of the projectile is horizontal and, because there is no acceleration in the horizontal direction, its speed is equal to the magnitude of the x-component of the initial velocity, $v_0 \cos\theta$.

$$0 \text{ J} = W_{\text{nonconservative}} = \Delta(\text{KE}+\text{PE}) = (\text{KE}+\text{PE})_{\text{f}} - (\text{KE}+\text{PE})_i = \left(\frac{m(v_0 \cos\theta)^2}{2} + mgy_{\text{f}} \right) - \left(\frac{mv_0^2}{2} + 0 \text{ J} \right)$$

$$\implies mgy_{\text{f}} = \frac{mv_0^2}{2} - \frac{mv_0^2 \cos^2\theta}{2} \implies y_{\text{f}} = \frac{v_0^2(1 - \cos^2\theta)}{2g} = \frac{v_0^2 \sin^2\theta}{2g}.$$

8.53

a) The appropriate gravitational potential energy is $\text{PE} = -\dfrac{GMm}{r}$. So the gravitational potential energies at points A and B are

$$\text{PE}_A = -\frac{GMm}{r_1} \quad \text{and} \quad \text{PE}_B = -\frac{GMm}{r_2}.$$

b) The orbital angular momentum is $\vec{L} = \vec{r} \times \vec{p}$. Its magnitude is $L = rp\sin\theta = rmv\sin\theta$, where θ is the angle between $\vec{r}$ and $\vec{p}$. At points A and B, the angle between $\vec{r}$ and $\vec{p}$ is $90°$. Hence, the magnitudes of the orbital angular momenta at A and B are

$$L_A = r_1 mv_1 \sin 90° = mv_1 r_1 \quad \text{and similarly} \quad L_B = mv_2 r_2.$$

The angular momentum is conserved, so

$$L_A = L_B \implies mv_1 r_1 = mv_2 r_2 \implies v_1 r_1 = v_2 r_2.$$

Note that this is the differential form of Kepler's law that "equal areas are swept out in equal times."

c) The conservative gravitational force of the planet is the only force acting on the satellite, so there is therefore zero work done by nonconservative forces. Therefore

$$0 \text{ J} = W_{\text{nonconservative}} = \Delta(\text{KE} + \text{PE}) = \Delta E.$$

Hence, the total mechanical energy of the satellite is the same at points A and B. For point A, write

$$E = \frac{mv_1^2}{2} + \left(-\frac{GMm}{r_1}\right) \implies$$

(1) $$mv_1^2 = 2E + 2\frac{GMm}{r_1}.$$

Similarly, at point B we have

(2) $$mv_2^2 = 2E + 2\frac{GMm}{r_2}.$$

From part b) we have $v_1 r_1 = v_2 r_2$ so $v_1 = \dfrac{r_2}{r_1}v_2$. Substitute this expression for v_1 into equation (1):

$$m\left(\frac{r_2}{r_1}v_2\right)^2 = 2E + 2\frac{GMm}{r_1} \implies \left(\frac{r_2}{r_1}\right)^2 mv_2^2 = 2E + 2\frac{GMm}{r_1}.$$

Now substitute the expression for mv_2^2 from equation (2):

$$\left(\frac{r_2}{r_1}\right)^2 \left(2E + 2\frac{GMm}{r_2}\right) = 2E + 2\frac{GMm}{r_1}$$

We now solve this equation for E:

$$\left(\frac{r_2}{r_1}\right)^2 \left(2E + 2\frac{GMm}{r_2}\right) = 2E + 2\frac{GMm}{r_1} \implies$$

$$\left(\frac{r_2^2}{r_1^2} - 1\right) E = \frac{GMm}{r_1} - \frac{GMmr_2}{r_1^2} \implies$$

$$\left(\frac{r_2^2 - r_1^2}{r_1^2}\right) E = \frac{GMmr_1 - GMmr_2}{r_1^2} \implies$$

$$E = \frac{GMm(r_1 - r_2)}{r_2^2 - r_1^2} = \frac{GMm(r_1 - r_2)}{(r_2 - r_1)(r_2 + r_1)} = -\frac{GMm}{r_2 + r_1}.$$

But $r_2 + r_1$ is the major axis of the ellipse and therefore twice the semimajor axis a. Hence

$$E = -\frac{GMm}{2a}.$$

Usually each orbit has its own unique energy. Cases like this one, however, in which several different orbits (or states) have the same energy, are called by the delightful name "degenerate," which is no reflection on their moral values! More specifically, we say an energy E "is degenerate" if its value does not uniquely determine the associated orbit. Although more commonly used in quantum mechanics, the concepts of degeneracy and perturbation theory actually originated in the theory of planetary orbits.

Finally, the central force $-kr^{-2}\hat{\mathbf{r}}$ conserves three quantities: the total energy E, the total angular momentum $\vec{\mathbf{L}}$, and the Runge-Lenz vector $\vec{\mathbf{A}}$ (not discussed in the text!), which points along the semimajor axis of the orbit and has magnitude ϵ, where ϵ is the eccentricity. Although energy and angular momentum are conserved by *any* central force, the Runge Lenz vector is *only* conserved by forces of the form $-kr^n\hat{\mathbf{r}}$, where n is an integer.

8.57   Take the expression for kinetic energy and differentiate it with respect to v.

$$\frac{d}{dv}\text{KE} = \frac{d}{dv}\left(\frac{mv^2}{2}\right) = mv.$$

Thus for small changes Δv in speed from an initial speed v_0, the corresponding change ΔKE in kinetic energy is given approximately by

$$\Delta\text{KE} \approx mv_0\Delta v.$$

Solving for Δv and using the CWE theorem, $W_{\text{total}} = \Delta\text{KE}$,

$$\Delta v \approx \frac{\Delta\text{KE}}{mv_0} = \frac{W_{\text{total}}}{mv_0}.$$

With the given data

$$\Delta v \approx \frac{1.00\text{ J}}{(10.0\text{ kg})(30.0\text{ m/s})} \approx 3.33 \times 10^{-3}\text{ m/s}.$$

8.61

a)   The velocity component is the time derivative of $x(t)$,

$$v_x(t) = \frac{d}{dt}x(t) = -A\omega\sin(\omega t).$$

b)

$$\begin{aligned}
\text{PE}_{\text{ave}} &= \frac{1}{T}\int_0^T \frac{kx(t)^2}{2}\,dt \\
&= \frac{1}{2T}\int_0^T k[A\cos(\omega t)]^2\,dt \\
&= \frac{kA^2}{2T\omega}\int_0^T \cos^2(\omega t)\,\omega dt \\
&= \frac{kA^2}{2T\omega}\left[\frac{\omega t}{2} + \frac{1}{4}\sin(2\omega t)\right]_0^T \\
&= \frac{kA^2}{4}.
\end{aligned}$$

c)

$$\text{KE}_{\text{ave}} = \frac{1}{T} \int_0^T \frac{mv(t)^2}{2}\, dt$$

$$= \frac{1}{2T} \int_0^T m[A\omega \sin(\omega t)]^2\, dt$$

$$= \frac{\omega m A^2}{2T} \int_0^T \sin^2(\omega t)\, \omega dt$$

$$= \frac{\omega m A^2}{2T} \left[\frac{\omega t}{2} - \frac{1}{4}\sin(2\omega t) \right]_0^T$$

$$= \frac{\omega^2 m A^2}{4}$$

d) Recall that the angular frequency of the oscillation is $\omega = \sqrt{\dfrac{k}{m}}$. Therefore, the average value of the kinetic energy is

$$\text{KE}_{\text{ave}} = \frac{\omega^2 m A^2}{4} = \frac{m A^2 k}{4m} = \frac{k A^2}{4} = \text{PE}_{\text{ave}}.$$

This is a special case of a more general result, called the *virial theorem*, which states that if $\vec{\mathbf{F}}$ can be written in the form $\vec{\mathbf{F}}(\vec{\mathbf{r}}) = (Ar^n)\hat{\mathbf{r}}$, then $\text{KE}_{\text{ave}} = \left(\dfrac{n+1}{2} \right) \text{PE}_{\text{ave}}$ (where the average is taken over one period).

8.65

a) The angular frequency of the oscillation is $\omega = \sqrt{\dfrac{k}{m}}$. The frequency ν of the oscillation is related to the angular frequency by $\nu = \dfrac{\omega}{2\pi}$. The period T is the inverse of the frequency, so

$$T = \frac{1}{\nu} = 2\pi\sqrt{\frac{m}{k}} \implies k = \frac{4\pi^2 m}{T^2} = \frac{4\pi^2 4.00 \text{ kg}}{(1.50 \text{ s})^2} = 70.2 \text{ N/m}.$$

b) The total mechanical energy E is related to the amplitude by

$$E = \frac{k A^2}{2} \implies A = \sqrt{\frac{2E}{k}} = \sqrt{\frac{2(12.0 \text{ J})}{70.2 \text{ N/m}}} = 0.585 \text{ m}.$$

8.69

a) The work done by the gravitational force on the mass is equal to the negative of the change in its gravitational potential energy.

$$W = -\Delta\text{PE} = -(\text{PE}_f - \text{PE}_i).$$

Take $\hat{\mathbf{j}}$ pointing upward with the origin on the ground. Then

$$W = -(mgy_f - mgy_i) = -mg(y_f - y_i) = -(2.00 \text{ kg})(9.81 \text{ m/s}^2)(0 \text{ m} - 100 \text{ m}) = 1.96 \times 10^3 \text{ J}.$$

To find the time of the fall, compute

$$y(t) = y_0 + v_{y0}t + a_y\frac{t^2}{2} = y_0 - g\frac{t^2}{2} = 100 \text{ m} - g\frac{t^2}{2}.$$

Impact occurs where $y = 0 \text{ m}$, so

$$0 \text{ m} = 100 \text{ m} - (9.81 \text{ m/s}^2)\frac{t^2}{2}.$$

Solve for t, taking the positive root:

$$t = 4.52 \text{ s}.$$

Therefore the length Δt of the time interval during which the work took place is 4.52 s. So the average power is

$$P_{\text{ave}} = \frac{W}{\Delta t} = \frac{1.96 \times 10^3 \text{ J}}{4.52 \text{ s}} = 434 \text{ W}.$$

b)   The velocity component of the mass the instant before impact is

$$v_y(t) = v_{y0} + a_y t = -gt.$$

When $t = 4.52$ s,

$$v_y = (-9.81 \text{ m/s}^2)(4.52 \text{ s}) = -44.3 \text{ m/s}.$$

Therefore, the velocity is $\vec{v} = v_y \hat{j} = -44.3 \text{ m/s}\hat{j}$. The gravitational force $\vec{F}_{\text{gravity}}$ is

$$\vec{F}_{\text{gravity}} = -mg\hat{j} = -(2.00 \text{ kg})(9.81 \text{ m/s}^2)\hat{j} = -(19.6 \text{ N})\hat{j}.$$

The instantaneous power P of this force the instant before impact is

$$P = \vec{F}_{\text{gravity}} \bullet \vec{v} = (-19.6 \text{ N})\hat{j} \bullet (-44.3 \text{ m/s})\hat{j} = 868 \text{ W}.$$

8.73

a)   Choose a coordinate system with $\hat{j}$ pointing up and $\hat{i}$ pointing to the right, in the direction of the student's horizontal velocity component. Choose the origin 4.0 m directly below the student's starting position, so her initial position vector is $(4.0 \text{ m})\hat{j}$.

b)   At the top of the slide the speed of the student is zero, so the kinetic energy is zero as well. Therefore the total energy is

$$E = 0 \text{ J} + mgy_i = (80.0 \text{ kg})(9.81 \text{ m/s}^2)(4.0 \text{ m}) = 3.1 \times 10^3 \text{ J}.$$

c)   Use the CWE theorem with the initial position at the top of the incline and the final position at the bottom of the incline before the rough ground. During the slide down the frictionless incline, the normal force does no work, while the work from the weight is accounted for by the potential energy term. Thus

$$0 \text{ J} = W_{\text{nonconservative}} = \Delta(\text{KE} + \text{PE}) = (\text{KE} + \text{PE})_f - (\text{KE} + \text{PE})_i = (\frac{mv^2}{2} + 0 \text{ J}) - (0 \text{ J} + mgy_i)$$

$$\implies v = \sqrt{2gy_i} = \sqrt{2(9.81 \text{ m/s}^2)(4.0 \text{ m})} = 8.9 \text{ m/s}.$$

d)   Use the CWE theorem with the initial position at the bottom of the incline before the rough ground and the final position at rest some distance ℓ along the rough ground. Along the horizontal section, the normal force has a magnitude equal to that of the weight. (This can be shown by applying Newton's second law to the vertical direction along which there is zero acceleration). The force of kinetic friction is a constant force along the straight, horizontal path, so its work is

$$W_{\text{nonconservative}} = \vec{f}_k \bullet \Delta \vec{r} = f_k \Delta r (\cos 180°) = -f_k \Delta r = -\mu_k mg\ell.$$

Therefore

$$W_{\text{nonconservative}} = \Delta(\text{KE} + \text{PE})$$

$$\implies -\mu_k mg\ell = (\text{KE} + \text{PE})_f - (\text{KE} + \text{PE})_i = (0 \text{ J} + 0 \text{ J}) - (\frac{mv^2}{2} + 0 \text{ J})$$

$$\implies \ell = \frac{v^2}{2\mu_k g} = \frac{(8.9 \text{ m/s})^2}{2(0.20)(9.81 \text{ m/s}^2)} = 20 \text{ m}.$$

e)   The normal force does no work since it is always perpendicular to any change in the position vector of the system.

f)   The work done by the force of kinetic friction was found in part d) to be

$$W_{\text{nonconservative}} = -\mu_k mg\ell = -(0.20)(80.0 \text{ kg})(9.81 \text{ m/s}^2)(20 \text{ m}) = -3.1 \times 10^3 \text{ J}$$

g)   The power of the kinetic friction force as the student encounters the rough ground is

$$P = \vec{\mathbf{f}}_k \bullet \vec{\mathbf{v}} = (-\mu_k mg\hat{\mathbf{i}}) \bullet (v_x \hat{\mathbf{i}}) = -\mu_k mgv_x = -(0.20)(80.0 \text{ kg})(9.81 \text{ m/s}^2)(8.9 \text{ m/s}) = -1.4 \times 10^3 \text{ W}.$$

Since the velocity of the student decreases over the rough ground, the instantaneous power of the force of kinetic friction becomes less negative — eventually becoming zero when the student stops moving.

8.77   Convert the final speed of the car from km/h to m/s.

$$v = 150 \text{ km/h} = (150 \text{ km/h})\left(\frac{10^3 \text{ m}}{\text{km}}\right)\left(\frac{\text{h}}{3600 \text{ s}}\right) = 41.7 \text{ m/s}.$$

According to the CWE theorem, the work done by the total force is equal to the change in the kinetic energy of the car, so

$$W_{\text{total}} = \Delta \text{KE} = \frac{mv_f^2}{2} - \frac{mv_i^2}{2} = \frac{m(v_f^2 - v_i^2)}{2} = \frac{(1.00 \times 10^3 \text{ kg})\left[(41.7 \text{ m/s})^2 - (0 \text{ m/s})^2\right]}{2} = 8.69 \times 10^5 \text{ J}.$$

The average power of the total force is

$$P_{\text{ave}} = \frac{W_{\text{total}}}{\Delta t} = \frac{8.69 \times 10^5 \text{ J}}{8.00 \text{ s}} = 1.09 \times 10^5 \text{ W} = 109 \text{ kW}.$$

8.81

a)   Since the SI units of the potential energy are joules, and those of distance are meters, the SI units for the constant C must be J/m.

b)   The x-component of the force acting on the system, here the only force component, is

$$F_x = -\frac{d}{dx}\text{PE} = \begin{cases} -C, & \text{for } x > 0 \text{ m}, \\ C & \text{for } x < 0 \text{ m}. \end{cases}$$

c)   A graph of the potential energy as a function of x appears below. The absolute value of the slope of each line is C.

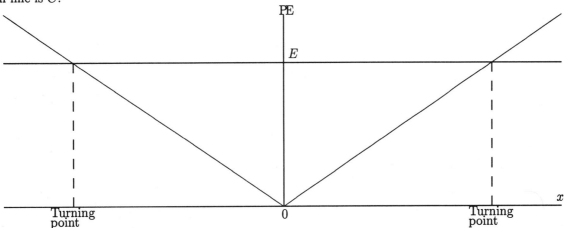

d)   A particular value E of the total mechanical energy also is shown in the graph above. The potential energy cannot be greater than the total mechanical energy, since that would imply a negative kinetic energy (and the kinetic energy is intrinsically positive). Thus, there are two turning points, as indicated on the graph above. Values of $|x|$ greater than the turning points are prohibited, since for these coordinates the potential energy is greater than the total mechanical energy.

e)   The motion of the particle is confined to coordinates x between the two turning points. The motion is oscillatory, but not simple harmonic oscillation since the potential energy function is not that of a simple harmonic oscillator — which has the form $\mathrm{PE} = \dfrac{kx^2}{2}$.

For positive values of x, the force is directed toward $-\hat{\mathbf{i}}$, and for negative values of x, the force is directed toward $\hat{\mathbf{i}}$. In this respect the force is similar to a Hooke's law force. But unlike the Hooke's law force, the given force has a constant magnitude for all values of x.

Chapter 9

Impulse, Momentum, and Collisions

9.1 Since the speed of the truck is given in (km/h) and you want your speed in (km/h), you do not need to convert the speeds to SI units. Equate the magnitude of the momentum of the dump truck with the magnitude of your momentum running at speed v:

$$p_{\text{truck}} = p_{\text{you}} \implies m_{\text{truck}}v_{\text{truck}} = m_{\text{you}}v \implies (2.0 \times 10^4 \text{ kg})(1.0 \text{ km/h}) = (60 \text{ kg})v \implies v = 3.3 \times 10^2 \text{ km/h},$$

which is about 200 mile/hour.

9.5 Convert the speed from (km/h) to (m/s):

$$1.00 \text{ km/h} = (1.00 \text{ km/h}) \left(\frac{10^3 \text{ m}}{\text{km}} \right) \left(\frac{\text{h}}{3600 \text{ s}} \right) = 0.278 \text{ m/s}.$$

Equate the magnitude of the momentum of the car with that of the student:

$$m_{\text{student}}v_{\text{student}} = m_{\text{car}}v_{\text{car}} \implies (60.0 \text{ kg})v_{\text{student}} = (1.00 \times 10^3 \text{ kg})(0.278 \text{ m/s}) \implies v_{\text{student}} = 4.63 \text{ m/s}.$$

Convert the speed to km/h if desired:

$$v_{\text{student}} = (4.63 \text{ m/s}) \left(\frac{\text{km}}{10^3 \text{ m}} \right) \left(\frac{3600 \text{ s}}{\text{h}} \right) = 16.7 \text{ km/h}.$$

9.9 Use the CWE theorem to determine the speed of the cockleshell the instant before impact. Choose a coordinate system with $\hat{\mathbf{j}}$ pointing up, $\hat{\mathbf{i}}$ pointing horizontally in the direction of the gull's flight, and origin at ground level directly below the release point.

The conservative gravitational force of the Earth is the only force acting on the shell during its fall. There are no nonconservative forces, so their work is zero. The CWE theorem becomes

$$0 \text{ J} = W_{\text{nonconservative}} = \Delta(\text{KE} + \text{PE}) = (\text{KE} + \text{PE})_{\text{f}} - (\text{KE} + \text{PE})_{\text{i}} = \left(\frac{mv_{\text{f}}^2}{2} + mg(0 \text{ m}) \right) - \left(\frac{mv_{\text{i}}^2}{2} + mgy_{\text{i}} \right)$$

$$\implies v_{\text{f}} = \sqrt{v_{\text{i}}^2 + 2gy_{\text{i}}} = \sqrt{(15.0 \text{ m/s})^2 + 2(9.81 \text{ m/s}^2)(20.0 \text{ m})} = 24.8 \text{ m/s}.$$

The magnitude of the momentum of the shell the instant before impact is

$$p = mv_{\text{f}} = (0.200 \text{ kg})(24.8 \text{ m/s}) = 4.96 \text{ kg·m/s}.$$

9.13

a) Choose a coordinate system with origin on the ground at the point where the egg lands (10 m below the release point), and with $\hat{\mathbf{i}}$ pointing up.

b)   During the fall, the conservative gravitational force of the Earth is the only force acting on the egg. Hence, there is zero work done by nonconservative forces and CWE theorem becomes

$$0 \text{ J} = W_{\text{nonconservative}} = \Delta(\text{KE} + \text{PE}) = (\text{KE} + \text{PE})_f - (\text{KE} + \text{PE})_i.$$

Thus, the total mechanical energy of the egg is conserved during the fall. The initial total mechanical energy is

$$E_i = \text{KE}_i + \text{PE}_i = 0 \text{ J} + mgy_i = (1.00 \text{ kg})(9.81 \text{ m/s}^2)(10.0 \text{ m}) = 98.1 \text{ J}.$$

Since the total mechanical energy is conserved during the fall, the total mechanical energy at a height of 5.0 m above the ground also is 98.1 J.

c)   Since the total force on the egg is constant (it is the weight of the egg), Newton's second law implies that the rate at which the momentum of the egg changes with time also is constant.

d)   The impulse imparted to the egg during its fall is

$$\vec{I} = \int_0^{\Delta t} \vec{F} \, dt = \vec{F}\Delta t,$$

since the force $\vec{F}$ is constant. We need to find the time t of the fall. With the coordinate system in part a), use the kinematic equation for motion with a constant acceleration:

$$y(t) = y_0 + v_{y0}t + a_y\frac{t^2}{2} = 10.0 \text{ m} + (0 \text{ m/s})t - (9.81 \text{ m/s}^2)\frac{t^2}{2}.$$

Impact is where $y = 0 \text{ m}$, so

$$0 \text{ m} = (10.0 \text{ m}) - (9.81 \text{ m/s}^2)\frac{t^2}{2} \implies t = 1.43 \text{ s}.$$

The interval of the fall is thus $\Delta t = t - 0 \text{ s} = 1.43 \text{ s}$, so the impulse is

$$I = (-mg\hat{j})\Delta t = (-1.00 \text{ kg})(9.81 \text{ m/s}^2)(1.43 \text{ s})\hat{j} = (-14.0 \text{ N·s})\hat{j}.$$

9.17

a)   The x-component of the impulse is the area under the graph of F_x versus t between the two times:

$$I_x = \frac{1}{2}(15.0 \text{ N})(2.00 \text{ s}) = 15.0 \text{ N·s} \implies \vec{I} = (15.0 \text{ N·s})\hat{i}.$$

b)   The average force times the time interval over which it acts is equal to the impulse:

$$\vec{F}_{\text{ave}}\Delta t = \vec{I} \implies F_{\text{ave}} = \frac{1}{\Delta t}\vec{I} = \left(\frac{15.0 \text{ N·s}}{2.00 \text{ s}}\right)\hat{i} = (7.50 \text{ N})\hat{i}.$$

c)   There is zero additional area under the curve between 2.00 s and 4.00 s, so the impulse is still

$$\vec{I} = (15.0 \text{ N·s})\hat{i}.$$

d)   The average force times the time interval over which it acts is equal to the impulse:

$$\vec{F}_{\text{ave}}\Delta t = \vec{I} \implies \vec{F}_{\text{ave}} = \frac{1}{\Delta t}\vec{I} = \left(\frac{15.00 \text{ N·s}}{4.00 \text{ s}}\right)\hat{i} = (3.75 \text{ N})\hat{i}.$$

9.21   To determine the impact speed of the painter, use the CWE theorem. Choose a coordinate system with $\hat{\mathbf{j}}$ pointing up and origin at the point of impact — 8.0 m directly below the painter's initial position.

During the unfortunate fall, the only force acting on the painter is the conservative gravitational force of the Earth. The CWE theorem then says that

$$0 \text{ J} = W_{\text{nonconservative}} = \Delta(\text{KE} + \text{PE}) = (\text{KE} + \text{PE})_f - (\text{KE} + \text{PE})_i = \left(\frac{mv^2}{2} + 0 \text{ J} \right) - (0 \text{ J} + mgy_i)$$

$$\implies v = \sqrt{2gy_i} = \sqrt{2(9.81 \text{ m/s}^2)(8.0 \text{ m})} = 13 \text{ m/s}.$$

So, since the painter is falling in the $-\hat{\mathbf{j}}$ direction, the painter's velocity at the time of impact is $(-13 \text{ m/s})\hat{\mathbf{j}}$. During the next 0.15 s the painter is "brought to rest;" so the average acceleration during the final 0.15 s is

$$\vec{\mathbf{a}}_{\text{ave}} = \frac{1}{\Delta t} \Delta \vec{\mathbf{v}} = \frac{1}{0.15 \text{ s}} \left((-13 \text{ m/s})\hat{\mathbf{j}} - \mathbf{0} \text{ m/s} \right) = (-87 \text{ m/s}^2)\hat{\mathbf{i}}.$$

The magnitude of this acceleration is 87 m/s^2.

9.25   Use Equation 9:15,

$$v_x = v_{\text{exhaust}} \ln \left(\frac{m_{\text{fuel}}}{m_{\text{payload}}} + 1 \right).$$

If the final speed v_y is equal to the speed of the exhaust gases, then

$$v_{\text{exhaust}} = v_{\text{exhaust}} \ln \left(\frac{m_{\text{fuel}}}{m_{\text{payload}}} + 1 \right)$$

$$\implies 1 = \ln \left(\frac{m_{\text{fuel}}}{m_{\text{payload}}} + 1 \right)$$

$$\implies e = \frac{m_{\text{fuel}}}{m_{\text{payload}}} + 1$$

$$\implies \frac{m_{\text{fuel}}}{m_{\text{payload}}} = e - 1 = 2.718 - 1 = 1.718.$$

9.29   Let the mass per unit length of the chain be λ, so

$$\lambda = \frac{m}{\ell}.$$

Consider the chain on the ground to be the system. When the top of the chain has fallen a distance y, the mass m of the chain on the ground is λy. The forces on this mass are its weight $\vec{\mathbf{w}}$, directed downward and the upward normal force $\vec{\mathbf{N}}$ of the ground. Take $\hat{\mathbf{j}}$ to point upward, and the origin at the point where the chain is released. Then Newton's second law applied in the vertical direction is

$$F_{y \text{ total}} = \frac{dp_y}{dt} \implies N - mg = \frac{d}{dt}(mv_y) = \left(\frac{dm}{dt} \right) v_y + m \left(\frac{dv_y}{dt} \right).$$

Since $m = \lambda y$, then $\dfrac{dm}{dt} = \lambda \dfrac{dy}{dt} = \lambda v_y$.

Also, the acceleration of the chain on the ground is zero so $\dfrac{dv_y}{dt} = 0 \text{ m/s}^2$. Therefore, Newton's second law reduces to

$$N - mg = \lambda v_y^2.$$

The chain falls with a constant acceleration, so use the kinematic equations for motion with a constant acceleration to determine the velocity component of the top of the chain when it has fallen a distance y. At time $t = 0 \text{ s}$, the chain is at rest so $v_{y0} = 0 \text{ m/s}$, and

$$v_y(t) = v_{y0} + a_y t = gt \implies N - mg = \lambda(gt)^2 = \lambda g^2 t^2.$$

Also, the distance that the top of the chain has fallen is

$$y(t) = y_0 + v_{y0}t + a_y \frac{t^2}{2} = g\frac{t^2}{2} \implies 2yg = g^2 t^2.$$

Newton's second law equation then is

$$N - mg = \lambda g^2 t^2 \implies N - \lambda yg = \lambda(2yg) \implies N = 3\lambda yg = 3\left(\frac{m}{\ell}\right)yg = \frac{3mgy}{\ell}.$$

9.33

a)   Conserve the total momentum before and after the collision:

$$\vec{\mathbf{P}}_{\text{total before}} = \vec{\mathbf{P}}_{\text{total after}} \implies m_{\text{car}}\vec{\mathbf{v}}_{\text{car}} = m_{\text{truck}}\vec{\mathbf{v}}_{\text{truck}}$$
$$\implies (1.00 \times 10^3 \text{ kg})\vec{\mathbf{v}}_{\text{car}} = (4.00 \times 10^3 \text{ kg})(2.00 \text{ m/s})\hat{\mathbf{i}} \implies \vec{\mathbf{v}}_{\text{car}} = (8.00 \text{ m/s})\hat{\mathbf{i}}$$

b)   To see if the collision is elastic, check the kinetic energy before and after the collision. Before the collision, the car is the only moving mass, and therefore

$$\text{KE}_{\text{before}} = \frac{m_{\text{car}}v_{\text{car}}^2}{2} = \frac{(1.00 \times 10^3 \text{ kg})(8.00 \text{ m/s})^2}{2} = 3.20 \times 10^4 \text{ J}.$$

After the crash, it is the truck that is moving, so

$$\text{KE}_{\text{after}} = \frac{m_{\text{truck}}v_{\text{truck}}^2}{2} = \frac{(4.00 \times 10^3 \text{ kg})(2.00 \text{ m/s})^2}{2} = 8.00 \times 10^3 \text{ J}.$$

The kinetic energy was not conserved in the collision, so it was not an elastic collision.

c)   The momentum of the car was not conserved; the force of the truck on the car changed the momentum of the car during the collision. Only the total momentum of all the masses involved in the collision is conserved before and after the collision, not the momentum of the individual masses.

9.37

a)   This is a stupid collision! It is also completely inelastic, since the cars stick together.

b)   Convert the speeds from km/h to m/s:

$$60.0 \text{ km/h} = (60.0 \text{ km/h})\left(\frac{10^3 \text{ m}}{\text{km}}\right)\left(\frac{\text{h}}{3600 \text{ s}}\right) = 16.7 \text{ m/s}.$$

$$40.0 \text{ km/h} = (40.0 \text{ km/h})\left(\frac{10^3 \text{ m}}{\text{km}}\right)\left(\frac{\text{h}}{3600 \text{ s}}\right) = 11.1 \text{ m/s}.$$

The total momentum before the collision is the vector sum of the cars' individual momenta:

$$\vec{\mathbf{P}}_{\text{before}} = (1.00 \times 10^3 \text{ kg})(16.7 \text{ m/s})\hat{\mathbf{i}} + (1.00 \times 10^3 \text{ kg})(11.1 \text{ m/s})\hat{\mathbf{j}}$$
$$= (16.7 \times 10^3 \text{ kg·m/s})\hat{\mathbf{i}} + (11.1 \times 10^3 \text{ kg·m/s})\hat{\mathbf{j}}.$$

c)   The total momentum after the collision must be the same, since the total momentum is conserved in a collision, so

$$\vec{\mathbf{P}}_{\text{after}} = \vec{\mathbf{P}}_{\text{before}} = (16.7 \times 10^3 \text{ kg·m/s})\hat{\mathbf{i}} + (11.1 \times 10^3 \text{ kg·m/s})\hat{\mathbf{j}}.$$

d)   Let $\vec{\mathbf{v}}$ be the velocity of the crumpled mass immediately after the collision. Then, since the total mass of the cars is 2.00×10^3 kg,

$$\vec{\mathbf{v}} = \frac{1}{2.00 \times 10^3}\vec{\mathbf{P}}_{\text{after}} = \frac{1}{2.00 \times 10^3}\left((16.7 \times 10^3 \text{ kg·m/s})\hat{\mathbf{i}} + (11.1 \times 10^3 \text{ kg·m/s})\hat{\mathbf{j}}\right)$$
$$= (8.35 \text{ m/s})\hat{\mathbf{i}} + (5.55 \text{ m/s})\hat{\mathbf{j}},$$

and the speed is $v = \sqrt{(8.35 \text{ m/s})^2 + (5.55 \text{ m/s})^2} = 10.0 \text{ m/s}.$

e)   The change in the kinetic energy is

$$\Delta KE = KE_{\text{total f}} - KE_{\text{total i}}$$

$$= \frac{(2.00 \times 10^3 \text{ kg})(10.0 \text{ m/s})^2}{2} - \left(\frac{(1.00 \times 10^3 \text{ kg})(16.7 \text{ m/s})^2}{2} + \frac{1.00 \times 10^3 \text{ kg})(11.1 \text{ m/s})^2}{2} \right)$$

$$= -1.01 \times 10^5 \text{ J}.$$

Where did all of this energy go? Probably most went into deforming the two car bodies. Some even went into making the load "crash" sound. Notice that about *half* of the initial kinetic energy went into these processes.

9.41

a)   Choose a coordinate system with origin at the point of impact, $\hat{\mathbf{i}}$ pointing east, and $\hat{\mathbf{j}}$ pointing north.

Let v be the speed of the (smashed) car-car system immediately after the collision. The total momentum of the system is conserved immediately before and after the collision, which means you can separately conserve the total x-component of the momentum and the total y-component of the momentum.

For the x-component we have

$p_{x \text{ total before}} = p_{x \text{ total after}}$

$$\implies (1.050 \times 10^3 \text{ kg})(22 \text{ m/s}) + (0.900 \times 10^3 \text{ kg})(18 \text{ m/s}) \cos 30° = (1.950 \times 10^3 \text{ kg})v_x$$

$$\implies 2.3 \times 10^4 \text{ kg·m/s} + 1.4 \times 10^4 \text{ kg·m/s} = (1.950 \times 10^3 \text{ kg})v_x$$

$$\implies v_x = 19 \text{ m/s}.$$

For the y-component we have

$p_{y \text{ total before}} = p_{y \text{ total after}}$

$$\implies -(0.900 \times 10^3 \text{ kg})(18 \text{ m/s}) \sin 30° = (1.950 \times 10^3 \text{ kg})v_y$$

$$\implies v_y = -4.2 \text{ m/s}.$$

Therefore, the velocity of the car-car system after collision is

$$\vec{v} = v_x\hat{\mathbf{i}} + v_y\hat{\mathbf{j}} = (19 \text{ m/s})\hat{\mathbf{i}} - (4.2 \text{ m/s})\hat{\mathbf{j}}.$$

Using the same precision for each component, the velocity is

$$v = (19 \text{ m/s})\hat{\mathbf{i}} - (4 \text{ m/s})\hat{\mathbf{j}}.$$

b)   The speed of the composite system after the collision is the magnitude of its velocity,

$$v = \sqrt{(19 \text{ m/s})^2 + (-4 \text{ m/s})^2} = 19 \text{ m/s}.$$

Therefore, the change in the kinetic energy is

$$\Delta KE = KE_f - KE_i = \frac{(1950 \text{ kg})(19 \text{ m/s})^2}{2} - \left(\frac{(1050 \text{ kg})(22 \text{ m/s})^2}{2} + \frac{(900 \text{ kg})(18 \text{ m/s})^2}{2} \right)$$

$$= 3.5 \times 10^5 \text{ J} - 2.5 \times 10^5 \text{ J} - 1.5 \times 10^5 \text{ J} = -0.5 \times 10^5 \text{ J} = -5 \times 10^4 \text{ J}.$$

9.45

a)   Let $\hat{\mathbf{i}}$ be in the direction of total momentum before and after the collision. (Both car and truck are going in the same direction). Let v_{x0} be your velocity component immediately before the collision. Conserve the total momentum before and after the collision:

$$\vec{\mathbf{P}}_{x \text{ total before}} = \vec{\mathbf{P}}_{x \text{ total after}} \implies (1.00 \times 10^3 \text{ kg})v_{x0} + (1.80 \times 10^3 \text{ kg})(15.0 \text{ m/s}) = (2.80 \times 10^3 \text{ kg})(20.0 \text{ m/s})$$

$$\implies v_{x0} = 29.0 \text{ m/s}.$$

You were not exceeding the legal speed limit.

b)   Let v_x be the velocity component of the composite system of three cars immediately after the collision. Conserve the total momentum before and after the collision:

$$\vec{\mathbf{P}}_{x \text{ total before}} = \vec{\mathbf{P}}_{x \text{ total after}} \implies (2.80 \times 10^3 \text{ kg})(20.0 \text{ m/s})\hat{\mathbf{i}} - (1.50 \times 10^3 \text{ kg})(40.0 \text{ m/s})\hat{\mathbf{i}} = (4.30 \times 10^3 \text{ kg})v_x\hat{\mathbf{i}}$$
$$\implies v_x = -0.930 \text{ m/s}.$$

The speed is the magnitude of the velocity vector, so the speed is 0.930 m/s.

9.49   Use the CWE theorem to find the speed v of m the instant before it collides with M. Take the initial position to be where m is released and the final position to be where it is about to collide with M. Choose a coordinate system with origin at the bottom of the inclined plane, $\hat{\mathbf{j}}$ pointing up, and $\hat{\mathbf{i}}$ pointing horizontally to the right.

   Since the slide is frictionless, there is no work done by the force of kinetic friction while m slides down the plane. There is work done by the force of kinetic friction along the rough, horizontal ground. Along the straight section of horizontal rough ground of length 0.20 m, the normal force is equal in magnitude to the weight of m. The force of kinetic friction is constant, so its work is

$$W_{\text{friction}} = \vec{\mathbf{f}}_k \bullet \Delta \vec{\mathbf{r}} = f_k \Delta r \cos 180° = (\mu_k mg)(0.20 \text{ m})(-1) = (-0.20 \text{ m})\mu_k mg.$$

This is the $W_{\text{nonconservative}}$ in the CWE theorem. The initial height of m is $y_i = (2.0 \text{ m}) \sin 30°$. So, applying the CWE theorem,

$$(-0.20 \text{ m})\mu_k mg = W_{\text{nonconservative}} = \Delta(\text{KE} + \text{PE}) = (\text{KE} + \text{PE})_f - (\text{KE} + \text{PE})_i$$
$$= \left(\frac{mv^2}{2} + 0 \text{ J}\right) - (0 \text{ J} + mg(2.0 \text{ m}) \sin 30°)$$
$$= \frac{mv^2}{2} - mg(2.0 \text{ m}) \sin 30° = \frac{mv^2}{2} - \frac{mg(2.0 \text{ m})}{2}.$$

Solve for v, and substitute the values for μ_k and g.

$$v = \sqrt{g(2.0 \text{ m}) - 2(0.20 \text{ m})\mu_k g} = \sqrt{(9.81 \text{ m/s}^2)(2.0 \text{ m}) - 2(0.20 \text{ m})(0.25)(9.81 \text{ m/s}^2)} = 4.4 \text{ m/s}.$$

Now let the system be m and M. Conserve the total momentum of the system before and after the collision. The particles stick together after the collision since it is completely inelastic. Let v' be their common speed after the collision. Then

$$\vec{\mathbf{P}}_{\text{total before}} = \vec{\mathbf{P}}_{\text{total after}} \implies mv = (m + M)v' \implies \frac{M}{m} = \frac{v}{v'} - 1 = \frac{4.4 \text{ m/s}}{0.25 \text{ m/s}} - 1 = 17.$$

9.53   Let the two particles have masses m_1 and m_2. Choose a coordinate system at rest on m_2, so the velocity of the second particle is 0 m/s before the collision. Let $\vec{\mathbf{v}}_1$ be the velocity of the first particle before impact, so v_1 is the relative speed of approach before impact. Let $\vec{\mathbf{v}}'_1$ and $\vec{\mathbf{v}}'_2$ be the velocities of particle one and two respectively after impact. So their relative "speed of recession" after impact is the magnitude of $\vec{\mathbf{v}}'_1 - \vec{\mathbf{v}}'_2$. We need to show that $\vec{\mathbf{v}}_1$ and $\vec{\mathbf{v}}'_1 - \vec{\mathbf{v}}'_2$ have the same magnitude. It is equivalent, and more convenient, to work with the squares of their magnitudes and show that

(1)
$$|\vec{\mathbf{v}}'_1 - \vec{\mathbf{v}}'_2|^2 = |\vec{\mathbf{v}}_1|^2.$$

   The collision is elastic so the total kinetic energy before and after the collision is conserved. Therefore, $\text{KE}_{\text{total before}} = \text{KE}_{\text{total after}}$. Thus $\dfrac{m_1 v_1^2}{2} = \dfrac{m_1 v_1'^2}{2} + \dfrac{m_2 v_2'^2}{2}$. This implies

(2)
$$v_1'^2 = v_1^2 - \frac{m_2}{m_1}v_2'^2.$$

Momentum also is conserved, so $\vec{\mathbf{p}}_{\text{total before}} = \vec{\mathbf{p}}_{\text{total after}}$. Therefore, $m_1\vec{\mathbf{v}}_1 = m_1\vec{\mathbf{v}}_1' + m_2\vec{\mathbf{v}}_2'$. Hence

$$(3) \qquad \vec{\mathbf{v}}_1 = \vec{\mathbf{v}}_1' + \frac{m_2}{m_1}\vec{\mathbf{v}}_2'.$$

Take the scalar product of this equation with itself to find the square of the relative speed of approach before the collision:

$$v_1^2 = \vec{\mathbf{v}}_1 \bullet \vec{\mathbf{v}}_1 = \left(\vec{\mathbf{v}}_1' + \frac{m_2}{m_1}\vec{\mathbf{v}}_2'\right) \bullet \left(\vec{\mathbf{v}}_1' + \frac{m_2}{m_1}\vec{\mathbf{v}}_2'\right) = v_1'^2 + 2\left(\frac{m_2}{m_1}\right)\vec{\mathbf{v}}_1' \bullet \vec{\mathbf{v}}_2' + \left(\frac{m_2}{m_1}\right)^2 v_2'^2.$$

Now solve this for the middle term on the far right hand side:

$$(4) \qquad 2\vec{\mathbf{v}}_1' \bullet \vec{\mathbf{v}}_2' = \frac{m_1}{m_2}v_1^2 - \frac{m_1}{m_2}v_1'^2 - \frac{m_2}{m_1}v_2'^2.$$

Compute

$$|\vec{\mathbf{v}}_1' - \vec{\mathbf{v}}_2'|^2 = (\vec{\mathbf{v}}_1' - \vec{\mathbf{v}}_2') \bullet (\vec{\mathbf{v}}_1' - \vec{\mathbf{v}}_2') = \vec{\mathbf{v}}_1' \bullet \vec{\mathbf{v}}_1' - 2\vec{\mathbf{v}}_1' \bullet \vec{\mathbf{v}}_2' + \vec{\mathbf{v}}_2' \bullet \vec{\mathbf{v}}_2' = v_1'^2 - 2\vec{\mathbf{v}}_1' \bullet \vec{\mathbf{v}}_2' + v_2'^2$$

then use (4) to substitute for the middle term on the far right,

$$|\vec{\mathbf{v}}_1' - \vec{\mathbf{v}}_2'|^2 = v_1'^2 - \frac{m_1}{m_2}v_1^2 + \frac{m_1}{m_2}v_1'^2 + \frac{m_2}{m_1}v_2'^2 + v_2'^2$$

Use equation (2) to substitute for $v_1'^2$ in the first and third terms on the right-hand side of this equation, and then simplify:

$$|\vec{\mathbf{v}}_1' - \vec{\mathbf{v}}_2'|^2 = v_1^2 - \frac{m_2}{m_1}v_2'^2 - \frac{m_1}{m_2}v_1^2 + \frac{m_1}{m_2}\left(v_1^2 - \frac{m_2}{m_1}v_2'^2\right) + \frac{m_2}{m_1}v_2'^2 + v_2'^2 = v_1^2.$$

9.57   The position vector of the center of mass is

$$\vec{\mathbf{r}}_{\text{CM}} = \frac{\sum_i m_i\vec{\mathbf{r}}_i}{\sum_i m_i} = \frac{m_1\vec{\mathbf{r}}_1 + m_2\vec{\mathbf{r}}_2 + m_3\vec{\mathbf{r}}_3}{m_1 + m_2 + m_3}$$

$$= \frac{(2.50\text{ kg})\mathbf{0}\text{ m} + (3.00\text{ kg})(0.400\text{ m})\hat{\mathbf{j}} + (1.50\text{ kg})[(-0.600\text{ m})\hat{\mathbf{i}} + (0.400\text{ m})\hat{\mathbf{j}}]}{2.50\text{ kg} + 3.00\text{ kg} + 1.50\text{ kg}}$$

$$= \frac{(-0.900\text{ kg·m})\hat{\mathbf{i}} + (1.80\text{ kg·m})\hat{\mathbf{j}})}{7.00\text{ kg}} = (-0.129\text{ m})\hat{\mathbf{i}} + (0.257\text{ m})\hat{\mathbf{j}}.$$

9.61   The volume of the hole is

$$\frac{4}{3}\pi\left(\frac{R}{4}\right)^3 = \frac{1}{64}\left(\frac{4\pi R^3}{3}\right).$$

Let m be the mass of the solid portion of the Earth (not including the hollow). Use the method of "judicious subtraction." Consider the given system to be composed of a solid sphere of mass

$$m_1 = \frac{65}{64}m.$$

minus the second sphere (Hell) of mass

$$m_2 = \frac{1}{64}m.$$

The position vector of the center of mass is

$$\vec{\mathbf{r}}_{\text{CM}} = \frac{m_1\vec{\mathbf{r}}_1 - m_2\vec{\mathbf{r}}_2}{m_1 - m_2} = \frac{\frac{65}{64}m\mathbf{0}\text{ m} - \frac{1}{64}m\left(\frac{R}{4}\right)\hat{\mathbf{i}}}{\frac{65}{64}m - \frac{1}{64}m} = -\frac{1}{256}R\hat{\mathbf{i}}.$$

9.65   The total momentum of the father-daughter system in the horizontal direction is conserved. Take $\hat{\imath}$ to be in the direction of motion of the daughter. Let v_x be the x-component of the velocity of the father after the parting. Then

$$p_{x \text{ total before}} = p_{x \text{ total after}} \implies 0 \text{ kg·m/s} + 0 \text{ kg·m/s} = (20 \text{ kg})(2.5 \text{ m/s}) + (70 \text{ kg})v_x \implies v_x = -0.71 \text{ m/s}.$$

Therefore the father moves in the direction opposite to that of his daughter with a speed of 0.71 m/s.

9.69

a)   Let $\hat{\imath}$ be in the direction of motion. The one-dimensional collision is completely inelastic. Conserve the total momentum before and after the collision (noting in this case that the momentum only has an x-component):

$$p_{x \text{ total before}} = p_{x \text{ total after}}$$
$$\implies (120 \times 10^3 \text{ kg})(2.0 \text{ m/s}) = (120 \times 10^3 \text{ kg} + 10 \times 10^3 \text{ kg})\, v_x \implies v_x = 1.8 \text{ m/s}.$$

The final speed is 1.8 m/s.

b)   The change in the kinetic energy is

$$\Delta \text{KE} = \text{KE}_f - \text{KE}_i = \frac{(130 \times 10^3 \text{ kg})(1.8 \text{ m/s})^2}{2} - \frac{(120 \times 10^3 \text{ kg})(2.0 \text{ m/s})^2}{2}$$
$$= 2.1 \times 10^5 \text{ J} - 2.4 \times 10^5 \text{ J} = -0.3 \times 10^5 \text{ J} = -3 \times 10^4 \text{ J}.$$

c)   The velocity of the center of mass is unchanged in the collision. Since the cars couple together, the velocity (and speed) of the center of mass is the same as that of the coupled system, which is 1.8 m/s.

Chapter 10

Spin and Orbital Motion

10.1   Take the origin at the fish. The position vector of the pelican is

$$\vec{r} = (-2.50 \text{ m})\hat{\mathbf{i}}.$$

The momentum of the pelican is

$$\vec{p} = m\vec{v} = (4.00 \text{ kg})(5.00 \text{ m/s})\hat{\mathbf{i}} = (20.0 \text{ kg·m/s})\hat{\mathbf{i}}.$$

The angular momentum of the pelican with respect to the fish is

$$\vec{L} = \vec{r} \times \vec{p} = (-2.50 \text{ m})\hat{\mathbf{i}} \times (20.0 \text{ kg·m/s})\hat{\mathbf{i}} = \mathbf{0} \text{ kg·m}^2/\text{s}.$$

The angular momentum is orbital angular momentum, since the pelican is treated as a particle and is not spinning.

10.5

a)   By definition the eccentricity is

$$\epsilon = \frac{c}{a},$$

where c is the distance from either focus to the center. So

$$c = \epsilon a.$$

The perihelion distance is the distance of the closest point on the ellipse to one of the two foci. It is

$$r_{\text{perihelion}} = a - c = a - \epsilon a = a(1 - \epsilon).$$

The aphelion distance is the distance of the farthest point on the ellipse to one of the two foci. It is

$$r_{\text{aphelion}} = a + c = a + \epsilon a = a(1 + \epsilon).$$

Therefore

$$\frac{r_{\text{aphelion}}}{r_{\text{perihelion}}} = \frac{a(1 + \epsilon)}{a(1 - \epsilon)} = \frac{1 + \epsilon}{1 - \epsilon}.$$

b)   Choose a coordinate system whose origin is the center of the ellipse. In this coordinate system the moment of inertia of the planet at each point is

$$I = mr^2 \implies \frac{I_{\text{aphelion}}}{I_{\text{perihelion}}} = \frac{mr^2_{\text{aphelion}}}{mr^2_{\text{perihelion}}} = \left(\frac{r_{\text{aphelion}}}{r_{\text{perihelion}}}\right)^2 = \left(\frac{1 + \epsilon}{1 - \epsilon}\right)^2.$$

10.9   Convert the angular speed from rev/s to rad/s:

$$1.00 \text{ rev/s} = (1.00 \text{ rev/s}) \left(\frac{2\pi \text{ rad}}{\text{rev}} \right) = 6.28 \text{ rad/s} \, .$$

In circular motion, the tangential speed v at a distance r from the center is related to the angular speed ω by $v = r\omega$, so $r = \frac{v}{\omega}$. We want v to be the speed of light, so

$$r = \frac{v}{\omega} = \frac{3.00 \times 10^8 \text{ m/s}}{6.28 \text{ rad/s}} = 4.78 \times 10^7 \text{ m} = 4.78 \times 10^4 \text{ km} \, ,$$

which is about thirty thousand miles!

10.13   The total moment of inertia is the sum of the moments of inertia of the various parts about the same axis, so

$$I = I_{\text{disk}} + I_{\text{rim}} + I_{\text{point particles}} = \frac{mR^2}{2} + mR^2 + 4 \left(\frac{m}{4} \right) \left(\frac{R}{2} \right)^2 = \frac{7}{4} mR^2 \, .$$

10.17   Call the moment of inertia about the given axis I_x. Then $I_x = I_y$, so

$$I_z = I_x + I_y = 2I_x \, .$$

But from Table 10.1, the moment of inertia of a thin hoop about an axis perpendicular to the plane of the hoop and passing through its center is mR^2. Here, this is I_z. So

$$mR^2 = 2I_x \implies I_x = \frac{mR^2}{2} \, .$$

10.21

a)   Convert the speed from km/h to m/s:

$$150 \text{ km/h} = (150 \text{ km/h}) \left(\frac{10^3 \text{ m}}{\text{km}} \right) \left(\frac{\text{h}}{3600 \text{ s}} \right) = 41.7 \text{ m/s} \, .$$

The kinetic energy of the center of mass is

$$\text{KE}_{\text{CM}} = \frac{mv^2}{2} = \frac{(1.500 \times 10^3 \text{ kg})(41.7 \text{ m/s})^2}{2} = 1.30 \times 10^6 \text{ J} \, .$$

From Table 10.1, the moment of inertia of each wheel (treated as a uniform disk) is

$$I = \frac{mr^2}{2} = \frac{(35.0 \text{ kg})(0.25 \text{ m})^2}{2} = 1.1 \text{ kg} \cdot \text{m}^2 \, .$$

Since each wheel is rolling without slipping, the translational speed of each wheel is

$$v = \omega r \implies \omega = \frac{v}{r} = \frac{41.7 \text{ m/s}}{0.25 \text{ m}} = 1.7 \times 10^2 \text{ rad/s} \, .$$

The kinetic energy of rotation associated with each wheel is

$$\text{KE}_{\text{rot}} = \frac{I\omega^2}{2} = \frac{(1.1 \text{ kg} \cdot \text{m}^2)(1.7 \times 10^2 \text{ rad/s})^2}{2} = 1.6 \times 10^4 \text{ J} \, .$$

There are four wheels, so the total kinetic energy of rotation is

$$\text{KE}_{\text{total rot}} = 6.4 \times 10^4 \text{ J} \, .$$

Therefore the total kinetic energy of the car is

$$\text{KE} = \text{KE}_{\text{CM}} + \text{KE}_{\text{total rot}} = 1.30 \times 10^6 + 6.4 \times 10^4 \text{ J} = 1.36 \times 10^6 \text{ J} \, .$$

b)   The fraction is

$$\frac{\text{KE}_{\text{total rot}}}{\text{KE}} = \frac{6.4 \times 10^4 \text{ J}}{1.36 \times 10^6 \text{ J}} = 0.047 \approx 5\%.$$

Thus, the rotational kinetic energy is not insignificant. This is why some sport cars have metal alloy wheels rather than steel wheels. By reducing the rotational kinetic energy, they can decrease the amount of energy required to get the wheels up to speed, and put this energy into increasing the translational kinetic energy — thereby increasing acceleration. The tradeoff is that steel wheels are stronger, and so are less likely to get bent by hitting ruts or curbs. Steel wheels also are less expensive than alloy wheels.

10.25

a)   The star is spherical in shape, so its moment of inertia is found from Table 10.1 to be

$$I = \frac{2mR^2}{5}.$$

The angular speed ω is related to the frequency ν of the spin by $\omega = 2\pi\nu$, and the spin ν is the reciprocal $\frac{1}{T}$ of the period T. Thus

$$\omega = \frac{2\pi}{T}.$$

Therefore

$$\text{KE} = \frac{I\omega^2}{2} = \frac{\left(\dfrac{2mR^2}{5}\right)\left(\dfrac{2\pi}{T}\right)^2}{2} = \frac{4\pi^2 mR^2}{5T^2}.$$

b)   Substitute the given data to evaluate the rotational kinetic energy.

$$\text{KE} = \frac{4\pi^2(2 \times 1.99 \times 10^{30} \text{ kg})(10 \times 10^3 \text{ m})^2}{5(3.33 \times 10^{-2})^2} = 2.83 \times 10^{42} \text{ J}.$$

c)   Differentiate the expression for the kinetic energy in part a) with respect to t using the chain rule.

$$\frac{d}{dt}\text{KE} = \left(\frac{d}{dT}\text{KE}\right)\left(\frac{dT}{dt}\right) = \left((-2)\frac{4\pi^2 mR^2}{5T^3}\right)\left(\frac{dT}{dt}\right) = \left(\frac{-2\text{KE}}{T}\right)\left(\frac{dT}{dt}\right).$$

d)   Convert the rate at which the period is changing from $\dfrac{\text{s}}{\text{d}}$ to $\dfrac{\text{s}}{\text{s}}$:

$$\frac{dT}{dt} = 3.65 \times 10^{-8}\frac{\text{s}}{\text{d}} = \left(3.65 \times 10^{-8}\frac{\text{s}}{\text{d}}\right)\left(\frac{\text{d}}{8.64 \times 10^4 \text{ s}}\right) = 4.22 \times 10^{-13}\frac{\text{s}}{\text{s}}.$$

The power loss is

$$\frac{d}{dt}\text{KE} = \left(\frac{-2\text{KE}}{T}\right)\left(\frac{dT}{dt}\right) = \left(\frac{-2(2.83 \times 10^{43} \text{ J})}{3.33 \times 10^{-2} \text{ s}}\right)\left(4.22 \times 10^{-13}\frac{\text{s}}{\text{s}}\right) = -7.17 \times 10^{31} \text{ W}.$$

The power output of the Sun is 3.83×10^{26} W, so the ratio of the power loss of the neutron star to the power output of the Sun is

$$\frac{7.17 \times 10^{31} \text{ W}}{3.83 \times 10^{26} \text{ W}} = 1.87 \times 10^5.$$

10.29   For each case use the parallel axis theorem:

$$I = I_{\text{CM}} + md^2.$$

The moments of inertia about the symmetry axis through the center of mass are found in Table 10.1.

a)

$$I = \frac{2}{5}mR^2 + mR^2 = \frac{7}{5}mR^2.$$

b)

$$I = \frac{1}{2}mR^2 + mR^2 = \frac{3}{2}mR^2.$$

c)

$$I = \frac{2}{3}mR^2 + mR^2 = \frac{5}{3}mR^2.$$

10.33

a)   The Earth makes one rotation in 23 h and 56 min, which is 8.616×10^4 s. Therefore, the angular speed of the spin of the Earth is

$$\omega_{\text{spin}} = \frac{2\pi \text{ rad}}{8.616 \times 10^4 \text{ s}} = 7.293 \times 10^{-5} \text{ rad/s}.$$

b)   The magnitude of the spin angular momentum of the Earth is

$$L_{\text{spin}} = I_{\text{CM}}\,\omega_{\text{spin}} = \frac{2mR^2}{5}\omega_{\text{spin}}$$

$$= \frac{2(5.98 \times 10^{24} \text{ kg})(6.37 \times 10^6 \text{ m})^2}{5}(7.293 \times 10^{-5} \text{ rad/s}) = 7.08 \times 10^{33} \text{ kg·m}^2/\text{s}.$$

c)   The angular momentum of the isolated Earth is conserved. The transport of such sediment increases the distance of that mass from the axis of rotation. Therefore, the moment of inertia of the Earth increases. To keep the angular momentum constant, the spin angular speed of the Earth decreases. Hence the length of the day (the rotational period of the Earth) increases, i.e., the "day" gets longer.

d)   The Earth completes one orbit of the Sun in approximately 365 d $= 3.16 \times 10^7$ s. The orbital angular speed of the Earth is, therefore,

$$\omega_{\text{orbital}} = \frac{2\pi \text{ rad}}{3.16 \times 10^7 \text{ s}} = 1.99 \times 10^{-7} \text{ rad/s}.$$

e)   For its orbital motion, consider the Earth to be a particle. Then its moment of inertia about the Sun is mr^2, where r is the radius of its orbital circle. The magnitude of the orbital angular momentum of the Earth is

$$L_{\text{orbital}} = mr^2\omega_{\text{orbital}} = (5.98 \times 10^{24} \text{ kg})(1.49 \times 10^{11} \text{ m})^2(1.99 \times 10^{-7} \text{ rad/s}) = 2.64 \times 10^{40} \text{ kg·m}^2/\text{s}.$$

f)   The ratio of the magnitudes of the orbital to spin angular momenta is

$$\frac{L_{\text{orbital}}}{L_{\text{spin}}} = \frac{2.64 \times 10^{40} \text{ kg·m}^2/\text{s}}{7.08 \times 10^{33} \text{ kg·m}^2/\text{s}} = 3.73 \times 10^6.$$

10.37 There is no torque on the spinning system as the fluid leaks out, so the total angular momentum of the system is conserved. Therefore, using the magnitudes, we have

$$L_{\text{before}} = L_{\text{after}} \implies I\omega_0 = I'\omega'.$$

The total moment of inertia initially is that of the shell and its fluid contents; afterwards, it is just that of the shell. Hence

$$\left(\frac{2}{3}mR^2 + \frac{2}{5}\frac{m}{2}R^2\right)\omega_0 = \frac{2}{3}mR^2\omega'. \implies \omega' = \frac{13}{10}\omega_0.$$

10.41 Use the coordinate system shown below. Note that each object drops the same vertical distance h.

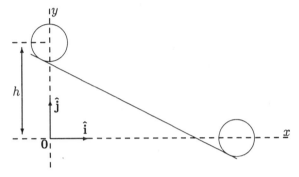

a) The force of static friction does no work on the objects as they roll down the incline, nor does the normal force, so the CWE theorem becomes

$$0 \text{ J} = W_{\text{nonconservative}} = \Delta(\text{KE} + \text{PE}) = (\text{KE} + \text{PE})_{\text{f}} - (\text{KE} + \text{PE})_{\text{i}} = \left(\frac{mv^2}{2} + \frac{I\omega^2}{2} + 0 \text{ J}\right) - (0 \text{ J} + mgh)$$

$$\implies mgh = \frac{mv^2}{2} + \frac{I\omega^2}{2} = \frac{mv^2}{2} + \frac{\beta mR^2\omega^2}{2}$$

where β is taken from Table 10:1. The rolling constraint implies $v = R\omega$, so

$$mgh = \frac{mv^2}{2} + \frac{\beta mR^2\left(\frac{v}{R}\right)^2}{2} \implies gh = \frac{(1+\beta)v^2}{2} \implies v = \sqrt{\frac{2gh}{1+\beta}}.$$

b) Note that the object with the highest value of β has the smallest value of v. Since each object has the same amount of total kinetic energy at the bottom of the incline, the one with the smallest v has the smallest kinetic energy associated with the motion of its center of mass, and correspondingly, the largest kinetic energy associated with rotation. The object with the smallest v also is the last one to reach the bottom of the incline. The ranking of the objects in terms of their rotational and translational kinetic energy corresponds to their values of β. From highest to lowest values of $\text{KE}_{\text{rotation}}$ they are: cylindrical shell, spherical shell, disk, and sphere. The ranking from highest to lowest KE_{CM} and speed is the reverse of this ranking.

c) The order of their arrival at the bottom of the incline proceeds from smallest to largest value of β: sphere, disk, spherical shell, cylindrical shell.

10.45

a) When the cable breaks, the forces on the beam are its weight and the force of the hinge on the beam. The force of the hinge on the beam produces zero torque about the hinge, since the line of action of the force passes through the point about which the torques are taken. Hence, the total torque is due only to the weight. The moment arm of the weight is 1.00 m. The magnitude of the torque is

$$\tau = mg(\text{moment arm}) = (200 \text{ kg})(9.81 \text{ m/s}^2)(1.00 \text{ m}) = 1.96 \times 10^3 \text{ N·m}.$$

b)   The moment of inertia of the beam about its end is found from the parallel axis theorem

$$I = I_{CM} + md^2.$$

Find I_{CM} from Table 10.1. Here $d = \dfrac{\ell}{2}$, so

$$I = \frac{m\ell^2}{12} + m\left(\frac{\ell}{2}\right)^2 = \frac{m\ell^2}{3} = \frac{(200 \text{ kg})(2.00 \text{ m})^2}{3} = 267 \text{ kg·m}^2.$$

The magnitude of the initial angular acceleration of the beam is found from

$$\alpha = \frac{\tau}{I} = \frac{1.96 \times 10^3 \text{ N·m}}{267 \text{ kg·m}^2} = 7.34 \text{ rad/s}^2.$$

c)   As the beam swings down, the moment arm of the weight decreases. Hence the magnitude of the torque decreases, as does the magnitude of the angular acceleration.

d)   Use the CWE theorem. The force of the hinge does no work. There are no nonconservative forces, so their work is zero. Choose a coordinate system with origin at the hinge, $\hat{i}$ horizontal to the right, and $\hat{j}$ pointing up.

The CWE theorem becomes

$$0 \text{ J} = W_{\text{nonconservative}} = \Delta(\text{KE} + \text{PE}) = (\text{KE} + \text{PE})_f - (\text{KE} + \text{PE})_i = \left(\frac{I\omega^2}{2} + \frac{-mg\ell}{2}\right) - (0 \text{ J} + 0 \text{ J}).$$

Solve for ω:

$$\omega = \sqrt{\frac{mg\ell}{I}} = \sqrt{\frac{mg\ell}{\left(\frac{m\ell^2}{3}\right)}} = \sqrt{\frac{3g}{\ell}} = \sqrt{\frac{3(9.81 \text{ m/s}^2)}{2.00 \text{ m}}} = 3.84 \text{ rad/s}.$$

10.49   Consider the planet of mass m to be a particle. Since the angular momentum is conserved, the magnitude (and direction) of the angular momentum is the same at both the perihelion and aphelion points.

$$L_{\text{perihelion}} = L_{\text{aphelion}} \implies I_{\text{perihelion}}\omega_{\text{perihelion}} = I_{\text{aphelion}}\omega_{\text{aphelion}} \implies \frac{\omega_{\text{aphelion}}}{\omega_{\text{perihelion}}} = \frac{I_{\text{perihilion}}}{I_{\text{aphelion}}}$$

In problem 10.5 part b) we found that

$$\frac{I_{\text{aphelion}}}{I_{\text{perihilion}}} = \left(\frac{1 + \epsilon}{1 - \epsilon}\right)^2.$$

Hence,

$$\frac{\omega_{\text{aphelion}}}{\omega_{\text{perihilion}}} = \frac{I_{\text{perihilion}}}{I_{\text{aphelion}}} = \left(\frac{1 - \epsilon}{1 + \epsilon}\right)^2.$$

10.53   Use the center of the Earth as origin, and let $\hat{i}$ point from the center of the Earth towards the center of the Moon. Let m_E and m_M be the masses of Earth and Moon respectively, and let $\vec{r}_E$ and $\vec{r}_M$ be their respective position vectors. Then $m_E = 5.98 \times 10^{24}$ kg, $m_M = 7.36 \times 10^{22}$ kg, $\vec{r}_E = \mathbf{0}$ m, and $\vec{r}_M = (3.84 \times 10^8 \text{ m})\hat{i}$.

a)   The center of mass is

$$\vec{r}_{CM} = \frac{m_E \vec{r}_E + m_M \vec{r}_M}{m_E + m_M} = \frac{\mathbf{0} \text{ kg·m} + (7.36 \times 10^{22} \text{ kg})(3.84 \times 10^8 \text{ m})\hat{i}}{5.98 \times 10^{24} \text{ kg} + 7.36 \times 10^{22} \text{ kg}} = (4.67 \times 10^6 \text{ m})\hat{i}.$$

b)   The angular speed of the orbital motion of the Earth and Moon about their center of mass is

$$\omega_{\text{orbital}} = \frac{2\pi \text{ rad}}{27.322 \text{ d}} = \frac{2\pi \text{ rad}}{27.322 \text{ d}} \left(\frac{d}{8.6400 \times 10^4 \text{ s}} \right) = 2.6617 \times 10^{-6} \text{ rad/s}.$$

The magnitude of the orbital angular momentum of the Moon about the center of mass of the Earth-Moon system is $L_{\text{Moon orbital}} = I\omega_{\text{orbital}}$. The moment of inertia is that of a point particle in orbital motion, so

$$L_{\text{Moon orbital}} = m_M (r_{\text{CM}} - r_M)^2 \omega_{\text{orbital}}$$
$$= (7.36 \times 10^{22} \text{ kg})(3.84 \times 10^8 \text{ m} - 4.67 \times 10^6)^2 (2.6617 \times 10^{-6} \text{ rad/s}) = 2.81 \times 10^{34} \text{ kg·m}^2/\text{s}.$$

c)   The magnitude of the orbital angular momentum of the Earth about the center of mass of the Earth-Moon system is $L_{\text{Earth orbital}} = I\omega_{\text{orbital}}$. The moment of inertia is that of a point particle in orbital motion, so

$$L_{\text{Earth orbital}} = m_E r_{\text{CM}}^2 \omega_{\text{orbital}}$$
$$= (5.98 \times 10^{24} \text{ kg})(4.67 \times 10^6 \text{ m})^2 (2.6617 \times 10^{-6} \text{ rad/s}) = (3.47 \times 10^{32}) \text{ kg·m}^2/\text{s}.$$

d)   The spin angular speed of the Moon is the same as its orbital angular speed since the Moon is in synchronous rotation. The magnitude of the spin angular momentum of the Moon is $L_{\text{Moon spin}} = I\omega_{\text{spin}}$. The moment of inertia is that of a sphere, so

$$L_{\text{Moon spin}} = \frac{2 m_M R_M^2}{5} \omega_{\text{spin}}$$
$$= \frac{2(7.36 \times 10^{22} \text{ kg})(1.74 \times 10^6 \text{ m})^2}{5} (2.6617 \times 10^{-6} \text{ rad/s}) = 2.37 \times 10^{29} \text{ kg·m}^2/\text{s}.$$

e)   The spin angular speed of the Earth is

$$\omega_{\text{Earth spin}} = \frac{2\pi \text{ rad}}{23.933 \text{ h}} = \left(\frac{2\pi \text{ rad}}{23.933 \text{ h}} \right) \left(\frac{h}{3600 \text{ s}} \right) = 7.2926 \times 10^{-5} \text{ rad/s}.$$

The magnitude of the spin angular momentum of the Earth is $L_{\text{Earth spin}} = I\omega_{\text{spin}}$. The moment of inertia is that of a sphere.

$$L_{\text{Earth spin}} = \frac{2 m_E R_E^2}{5} \omega_{\text{Earth spin}}$$
$$= \frac{2(5.98 \times 10^{24} \text{ kg})(6.37 \times 10^6 \text{ m})^2}{5} (7.2926 \times 10^{-5} \text{ rad/s}) = 7.08 \times 10^{33} \text{ kg·m}^2/\text{s}.$$

10.57

a)   The greater the amount of mass and the farther away it is from the axis, the greater the moment of inertia. Since the shell has more mass farther from the axis than the solid sphere with equal total mass and radius, the moment of inertia of the spherical shell is greater than that of the solid sphere.

b)   Let h be the height of the plane, m the mass of the object, R its radius, v its speed at the bottom of the plane, and ω its angular speed at the bottom of the plane. Then v and ω are related by $v = R\omega$ so

(1)
$$\omega = \frac{v}{R}.$$

In rolling the object down the inclined plane, there is no work done by either the normal force or the force of static friction. Hence, the CWE theorem indicates the total mechanical energy is conserved. At the top of the plane it has mgh potential energy and 0 J kinetic energy. At the bottom of the plane, it has 0 J potential energy and its kinetic energy is $\frac{mv^2}{2} + \frac{I\omega^2}{2}$. Therefore, by the CWE theorem,

(2)
$$mgh = \frac{mv^2}{2} + \frac{I\omega^2}{2}.$$

Now use (1) to substitute for ω in (2), and solve for v:

$$mgh = \frac{mv^2}{2} + \frac{I\left(\frac{v}{R}\right)^2}{2} \implies v^2 = \frac{2mghR^2}{mR^2 + I} \implies v = \sqrt{\frac{2mghR^2}{mR^2 + I}}.$$

From this equation we see that the bigger the moment of inertia I, the smaller the speed v. Therefore, the object with the greater moment of inertia will arrive at the bottom of the plane last. The spherical shell lags behind the solid sphere.

10.61 The lazy-susan rotates in the opposite sense of that of the mouse, or counterclockwise. Let v be the speed of the mouse on the rim of the lazy-susan measured with respect to the ground; let v' be the speed of a particle on the rim of the lazy-susan, again measured with respect to the ground. The speed v_0 of the mouse with respect to the lazy-susan is

$$v_0 = v + v'.$$

The angular speed ω of the mouse is related to its speed v by

$$v = \omega R.$$

The angular speed ω' of the lazy-susan is related to its speed v' by

$$v' = \omega' R.$$

The angular momentum of the entire system is conserved. The initial angular momentum of the system is zero, so the vector sum of the angular momentum of the mouse and that of the lazy-susan must sum to zero. Hence the mouse and lazy-susan have angular momenta L and L' of equal magnitude and opposite direction. Equate the magnitudes:

$$I'\omega' = I\omega \implies \frac{M}{2}R^2\omega' = mR^2\omega \implies M\omega' = 2m\frac{v}{R} = 2m\frac{v_0 - v'}{R} = 2m\left(\frac{v_0}{R} - \omega'\right) \implies \omega' = \left(\frac{1}{1 + \frac{M}{2m}}\right)\frac{v_0}{R}.$$

10.65

a) The spin angular speed of the Sun is

$$\omega_{\text{spin}} = \frac{2\pi \text{ rad}}{27 \text{ d}} = \left(\frac{2\pi \text{ rad}}{27 \text{ d}}\right)\left(\frac{d}{8.6400 \times 10^4 \text{ s}}\right) = 2.7 \times 10^{-6} \text{ rad/s}.$$

The magnitude of the spin angular momentum of the Sun is

$$L_{\text{spin}} = I_{\text{CM}}\omega_{\text{spin}} = \frac{2mr^2}{5}\omega_{\text{spin}} = \frac{2(1.99 \times 10^{30} \text{ kg})(6.96 \times 10^8 \text{ m})^2}{5}(2.7 \times 10^{-6} \text{ rad/s}) = 1.0 \times 10^{42} \text{ kg·m}^2/\text{s}.$$

b) The spin angular momentum of the Sun is 2% of the total angular momentum of the solar system, so

$$0.02L_{\text{solar system}} = 1.0 \times 10^{42} \text{ kg·m}^2/\text{s} \implies L_{\text{solar system}} = 5 \times 10^{43} \text{ kg·m}^2/\text{s}.$$

c) Suppose the Sun had the total angular momentum of the solar system. Then

$$L = I_{\text{CM}}\omega_{\text{spin}} = \frac{2mr^2}{5}\omega_{\text{spin}}$$

$$\implies 5 \times 10^{43} \text{ kg·m}^2/\text{s} = \frac{2(1.99 \times 10^{30} \text{ kg})(6.96 \times 10^8 \text{ m})^2}{5}\omega_{\text{spin}}$$

$$\implies \omega_{\text{spin}} = 1 \times 10^{-4} \text{ rad/s} = \frac{2\pi}{T}$$

$$\implies T = \frac{2\pi}{1 \times 10^{-4} \text{ rad/s}} = 6 \times 10^4 \text{ s} = 0.7 \text{ d}.$$

10.69 The first phase of the motion is the time t during which the falling mass m speeds up the disk from rest. The forces on the falling mass are its weight and the force of the cord on it. Choose the coordinate system with origin at the point where m is released, and with $\hat{\mathbf{j}}$ pointing down. Apply Newton's second law to this mass, letting $a = a_y$:

$$F_{y\text{ total}} = ma \implies mg - T = ma$$

so

(1)
$$T = m(g - a).$$

Use the kinematic equation for one-dimensional motion with a constant acceleration,

$$y(t) = y_0 + v_{y0}t + a_y\frac{t^2}{2} = 0\text{ m} + (0\text{ m/s})t + a\frac{t^2}{2} = a\frac{t^2}{2}.$$

When m hits the floor we have $h = a\dfrac{t^2}{2}$, so

(2)
$$a = \frac{2h}{t^2}.$$

The disk experiences a constant angular acceleration during the fall of m. Use the kinematic equation for motion with a constant angular acceleration to determine the angular speed of the disk when m hits the floor. Since the disk begins at rest, $\omega_z(t) = \omega_{z0} + \alpha_z t$. So, since $\omega_{z0} = 0\text{ rad/s}$,

(3)
$$\omega_z(t) = \alpha_z t.$$

Since the string rolls off the hub, the magnitude of the acceleration of m and the magnitude of the angular acceleration of the disk are related by the rolling constraint.

(4)
$$a = r\alpha_z \implies \alpha_z = \frac{a}{r}.$$

Substitute this for α_z in equation (3), and then substitute for a from equation (2). The angular speed of the disk when m hits the floor then is

(5)
$$\omega = \frac{a}{r}t = \frac{2h}{rt}.$$

The torque on the disk is provided by the force of the cord on it and the frictional torque of magnitude τ_f in the opposite direction. The moment arm of the force of the cord on the disk is the radius r of the hub. Thus

$$\tau_{\text{total}} = I\alpha \implies Tr - \tau_f = I\alpha_z.$$

Use equations (1), (2), and (4) to write this as

(6)
$$m(g - a)r - \tau_f = I\left(\frac{2h}{rt^2}\right).$$

Now examine the motion of the disk after m hits the floor. The frictional torque provides the angular acceleration to slow the disk to a stop. Let α' be the magnitude of this angular acceleration. During this phase of the motion, use the kinematic equations for motion with a constant angular acceleration. After a time t' we have (since the initial angular speed is given by equation (5))

$$0\text{ rad/s} = \frac{2h}{rt} + (-\alpha')t' \implies \alpha' = \frac{2h}{rtt'}.$$

The torque on the disk is only the frictional torque during the time t', so

$$-\tau_f = I(-\alpha') \implies \tau_f = I\left(\frac{2h}{rtt'}\right).$$

Use this expression for the frictional torque in equation (6):

$$m(g-a)r - I\frac{2h}{rtt'} = I\frac{2h}{rt^2} \implies I\frac{2h}{rt^2}\left(1 + \frac{t}{t'}\right) = mr(g-a).$$

Substitute for a from equation (2).

$$I = \frac{\frac{rt^2}{2h}mr\left(g - \frac{2h}{t^2}\right)}{1 + \frac{t}{t'}} = \frac{mr^2\left(\frac{gt^2}{2h} - 1\right)}{1 + \frac{t}{t'}}.$$

10.73   The forces on the ladder are its weight $\vec{w}$, the normal force $\vec{N}_1$ of the vertical, frictionless wall on the ladder, the normal force $\vec{N}_2$ of the horizontal surface, and the force of static friction $\vec{f}_s$. The second law force diagram and an appropriate coordinate choice are shown below.

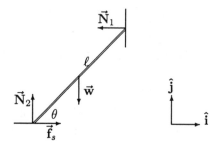

When the ladder is ready to slip, the force of static friction has its maximum magnitude

(1) $$f_{s\text{ max}} = \mu_s N_2.$$

Since the ladder is equilibrium, the sum of the forces along each coordinate axes must be zero. Therefore,

$$F_{x\text{ total}} = 0\text{ N} \implies f_{s\text{ max}} - N_1 = 0\text{ N}$$

so

(2) $$f_{s\text{ max}} = N_1.$$

Also

$$F_{y\text{ total}} = 0\text{ N} \implies N_2 - mg = 0\text{ N}$$

so

(3) $$N_2 = mg.$$

Since the ladder is in equilibrium, the sum of the torques taken about any point must be zero. Take torques about the base of the ladder. The forces $\vec{N}_2$ and $\vec{f}_{s\text{ max}}$ each produce zero torque about this point, since their lines of action pass through the point. Set the sum of the remaining torques to zero, so

$$0\text{ N·m} = \left(mg\frac{\ell}{2}\cos\theta\right)(-\hat{\mathbf{k}}) + (N_1\ell\sin\theta)\hat{\mathbf{k}} \implies N_1 = \frac{mg\cos\theta}{2\sin\theta} = \frac{mg}{2}\cot\theta.$$

Use equation (2) for N_1

$$f_{s\ max} = \frac{mg}{2}\cot\theta,$$

and then use equations (1) and (3) to find θ:

$$\mu_s mg = \frac{mg}{2}\cot\theta \implies \cot\theta = 2\mu_s \implies \tan\theta = \frac{1}{2\mu_s} = \frac{1}{2\times 0.30} = 1.70 \implies \theta = 59°.$$

So, at any angle less than 59°, the ladder will slip.

10.77 Choose a coordinate system with origin at the edge of the roof, $\hat{\mathbf{i}}$ pointing horizontally to the right, and $\hat{\mathbf{j}}$ pointing up. Let x be the maximum distance you can walk from the edge of the building without tipping the plank. When the plank is ready to rotate about the origin (the edge of the roof), the forces on the system are:

1. the weight $\vec{\mathbf{w}}_1$ of the plank, directed downward. This may be thought of as being applied at the center of mass of the plank, at the point $(-0.50\text{ m})\hat{\mathbf{i}}$;

2. your weight $\vec{\mathbf{w}}_2$, directed downward. This is applied at the point $x\hat{\mathbf{i}}$;

3. the force $\vec{\mathbf{F}}$ directed upward of the edge of the building on the plank. This is applied at the origin.

Take torques about the origin. Then force $\vec{\mathbf{F}}$ produces zero torque, since its line of action passes through the origin. Since the system is in equilibrium, the total torque on the system is zero, so

$$\vec{\tau}_{total} = \mathbf{0} \implies (100\text{ kg})(9.81\text{ m/s}^2)(0.50\text{ m}))\hat{\mathbf{k}} + (70.0\text{ kg})(9.81\text{ m/s}^2)x(-\hat{\mathbf{k}}) = \mathbf{0}\text{ N·m}$$

$$\implies (100\text{ kg})(9.81\text{ m/s}^2)(0.50\text{ m})) - (70.0\text{ kg})(9.81\text{ m/s}^2)x = 0\text{ N·m} \implies x = 0.71\text{ m}.$$

Chapter 11

Solids and Fluids

11.1 Young's modulus is

$$E = \frac{\text{stress}}{\text{strain}} = \frac{\frac{F}{A}}{\frac{\Delta\ell}{\ell}} \implies F = \frac{EA\Delta\ell}{\ell}.$$

The magnitude of the applied force on the wire is

$$mg = \frac{EA\Delta l}{\ell} \implies m = \frac{EA\Delta\ell}{g\ell}.$$

Substitute for the circular cross sectional area $A = \pi r^2$ of the wire, so

$$m = \frac{E\pi r^2 \Delta\ell}{g\ell}.$$

Use the numerical value of Young's modulus for steel from Table 11.1.

$$m = \frac{(20 \times 10^{10}\ \text{N/m}^2)\,\pi\,\dfrac{(1.0 \times 10^{-3}\ \text{m})^2}{2}(1.0 \times 10^{-3}\ \text{m})}{(9.81\ \text{m/s}^2)(2.00\ \text{m})} = 8.0\ \text{kg}.$$

11.5 Young's modulus E is

$$E = \frac{\text{stress}}{\text{strain}} = \frac{\frac{F}{A}}{\frac{\Delta\ell}{\ell}} = \frac{F\ell}{A\Delta\ell}.$$

The magnitude of the applied force on the line is the magnitude of the weight of the student. The area is the cross sectional area of the cord. Hence

$$E = \frac{mg\ell}{\pi r^2 \Delta\ell} = \frac{(80\ \text{kg})(9.81\ \text{m/s}^2)(2.00\ \text{m})}{\pi(4.0 \times 10^{-3}\ \text{m})^2(2.50\ \text{m})} = 1.2 \times 10^7\ \text{N/m}^2.$$

11.9 Young's modulus is

$$E = \frac{\text{stress}}{\text{strain}} = \frac{\frac{F}{A}}{\frac{\Delta\ell}{\ell}} = \frac{F\ell}{A\Delta\ell} \implies F = \frac{EA\Delta\ell}{\ell}.$$

Write the change in length as x instead of $\Delta\ell$. Then the applied force on the wire is

$$F = \frac{EAx}{\ell} = kx,$$

where $k \equiv \dfrac{EA}{\ell}$, the Hooke's law spring constant for the wire. The weight of m stretches the wire. The force of m on the wire and the force of the wire on m are a Newton's third law force pair, so the force on m is

$$\vec{F}_{\text{on } m} = -kx\hat{\mathbf{i}},$$

where $\hat{\mathbf{i}}$ points down. This is a Hooke's law force. Hence, from Chapter 7, the angular frequency ω of the oscillation of m is

$$\omega = \sqrt{\frac{k}{m}} = \sqrt{\frac{EA}{\ell m}}.$$

The frequency ν is related to the angular frequency ω by

$$\nu = \frac{\omega}{2\pi} = \frac{1}{2\pi}\sqrt{\frac{EA}{\ell m}}.$$

The cross sectional area of the wire is

$$A = \pi r^2 = \pi \left(\frac{d}{2}\right)^2.$$

Hence, the frequency is

$$\nu = \frac{1}{2\pi}\sqrt{\frac{E\pi\left(\dfrac{d}{2}\right)^2}{\ell m}} = \sqrt{\frac{Ed^2}{16\pi\ell m}}.$$

11.13   The pressure you exert on the floor is

$$\frac{\text{magnitude of weight}}{\text{area of feet}} = \frac{(60.0 \text{ kg})(9.81 \text{ m/s}^2)}{4.0 \times 10^{-2} \text{ m}^2} = 1.5 \times 10^4 \text{ Pa}.$$

Choose a coordinate system with $\hat{\mathbf{j}}$ pointing up and origin on the ground, and let y be the height of the equivalent column of water. The absolute pressure at the bottom of the water column is P_0, since the origin is there. The pressure P at the top of the water column is atmospheric pressure. The gauge pressure at the bottom of the water column is $P_0 - P = \rho g y$. So

$$1.5 \times 10^4 \text{ Pa} = \rho g y = (1.00 \times 10^3 \text{ kg/m}^3)(9.81 \text{ m/s}^2)y \implies y = 1.5 \text{ m}.$$

Note that this answer is roughly consistent with the fact that although the human body is mainly water, it still floats — and is therefore slightly less dense than water.

11.17

a)   Choose a coordinate system with $\hat{\mathbf{j}}$ pointing up and origin at ground level. Let y be the height of the hypothetical atmosphere. The pressure at the origin is $P_0 = 1.00$ atm $= 1.01 \times 10^5$ Pa. The pressure P at height y is 0 Pa. The pressures are related by

$$P(y) = P_0 - \rho g y \implies 0 \text{ Pa} = 1.01 \times 10^5 \text{ Pa} - (1.30 \text{ kg/m}^3)(9.81 \text{ m/s}^2)y \implies y = 7.92 \times 10^3 \text{ m}.$$

b)   The height of the hypothetical uniform atmosphere is 7.92 km, while the height of Everest is 8.85 km. Therefore, the top of the mountain is 8.85 km $-$ 7.92 km $=$ 0.93 km above the top of the hypothetical atmosphere.

11.21

a) The pistons are at the same elevation, so the pressures at each piston are the same.

b)

$$P_{\text{input}} = P_{\text{output}} \implies \frac{F_{\text{input}}}{A_{\text{input}}} = \frac{F_{\text{output}}}{A_{\text{output}}} \implies \frac{F_{\text{output}}}{F_{\text{input}}} = \frac{A_{\text{output}}}{A_{\text{input}}} = \frac{A_{\text{output}}}{\frac{1}{10}A_{\text{output}}} = 10.$$

11.25 The arrangement and a convenient coordinate system are the same as in Figure 11:16 in Example 11:3 of the text. Note that $\hat{\mathbf{j}}$ points up. The pressure as a function of height is given by

$$P(y) = P_0 - \rho g y.$$

The pressure at the top of the column is 0 Pa (a vacuum), while that at the reference level is atmospheric pressure. Hence

$$0\,\text{Pa} = 1.01 \times 10^5\,\text{Pa} - (1.00 \times 10^3)(\,\text{kg/m}^3)(9.81\,\text{m/s}^2)y \implies y = 10.3\,\text{m}.$$

11.29

a) The forces on the block are:

1. its weight $\vec{\mathbf{w}}$ pointing down; and

2. the buoyant force $\vec{\mathbf{F}}_{\text{buoyant}}$ pointing up.

The second law force diagram is shown below.

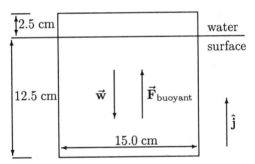

b) Since the block is in equilibrium, the total force on it must be zero. So

$$0\,\text{N} = F_{y\,\text{total}} = F_{\text{buoyant}} - w_y = F_{\text{buoyant}} - mg \implies F_{\text{buoyant}} = mg$$
$$\implies \rho_{\text{fluid}}V_{\text{block submerged}}g = \rho_{\text{wood}}V_{\text{block}}g$$
$$\implies \rho_{\text{wood}} = \rho_{\text{fluid}}\frac{V_{\text{block submerged}}}{V_{\text{block}}}$$
$$= (1.00 \times 10^3\,\text{kg/m}^3)\frac{(0.150\,\text{m})^2(0.125\,\text{m})}{(0.150)^3} = 0.833 \times 10^3\,\text{kg/m}^3.$$

11.33 Consider the ice cube and olive to be the system. The forces on the system are:

1. the weight $\vec{\mathbf{w}}_{\text{ice}}$ of the ice cube, directed down;

2. the weight $\vec{\mathbf{w}}_{\text{olive}}$ of the olive, directed up; and

3. the buoyant force $\vec{\mathbf{F}}_{\text{buoyant}}$ on the ice cube.

Choose a coordinate system with $\hat{j}$ directed up. The system is in equilibrium, so the sum of the forces acting on it is zero. Therefore

$$0 \text{ N} = F_{\text{total}} = F_{\text{buoyant}} - W_{\text{ice}} - W_{\text{olive}} = F_{\text{buoyant}} - m_{\text{ice}}g - m_{\text{olive}}g$$
$$= \rho_{\text{fluid}}V_{\text{ice}}g - \rho_{\text{ice}}V_{\text{ice}}g - m_{\text{olive}}g \implies m_{\text{olive}} = (\rho_{\text{fluid}} - \rho_{\text{ice}})V_{\text{ice}}.$$

Table 1.7 on page 12 of Chapter 1 gives the density of ice as $0.917 \times 10^3 \text{ kg/m}^3$. Therefore,

$$m_{\text{olive}} = (1.00 \times 10^3 \text{ kg/m}^3 - 0.917 \times 10^3 \text{ kg/m}^3)(3.0 \times 10^{-2} \text{ m})^3$$
$$= (0.08 \times 10^3 \text{ kg/m}^3)(27 \times 10^{-6} \text{ m}^3) = 2 \times 10^{-3} \text{ kg}.$$

11.37

a)   The volume of the barge is

$$V_{\text{barge}} = (1.00 \text{ m})(40.0 \text{ m}^2) = 40.0 \text{ m}^3.$$

The maximum magnitude of the buoyant force occurs when the barge is barely floating with its top at the same level as the surface of the water. In this case the buoyant force has magnitude

$$F_{\text{buoyant}} = \rho_{\text{water}}V_{\text{barge submerged}}g = (1.00 \times 10^3 \text{ kg/m}^3)(40.0 \text{ m}^3)(9.81 \text{ m/s}^2) = 3.92 \times 10^5 \text{ N}.$$

The magnitude of the weight of the empty barge is

$$W_{\text{empty barge}} = m_{\text{barge}}g = (1.00 \times 10^4 \text{ kg})(9.81 \text{ m/s}^2) = 9.81 \times 10^4 \text{ N}.$$

The magnitude of the weight of the barge is significantly less than the maximum value of the buoyant force. Therefore the barge will float. It will lower itself in the water until the magnitude of the buoyant force equals the magnitude of the weight.

b)   When the top of the barge is level with the water surface, the buoyant force has its maximum magnitude and the total force on the system is zero. Therefore,

$$0 \text{ N} = F_{\text{total}} = F_{\text{max buoyant}} - (m_{\text{barge}} + m_{\text{cargo}})g \implies F_{\text{max buoyant}} = (m_{\text{barge}} + m_{\text{cargo}})g$$
$$\implies (3.92 \times 10^5 \text{ N}) = (1.00 \times 10^4 \text{ kg} + m_{\text{cargo}})(9.81 \text{ m/s}^2) \implies m_{\text{cargo}} = 3.00 \times 10^4 \text{ kg}.$$

11.41

a)   and

b)   The forces acting on the cork are:

   1. its weight $\vec{\mathbf{w}}$, pointing down;

   2. the buoyant force $\vec{\mathbf{F}}_{\text{buoyant}}$, pointing up.

Here's the picture:

When the cork is floating and in equilibrium, the total force on it must be zero, so

$$0 \text{ N} = F_{\text{buoyant}} - W = F_{\text{buoyant}} - m_{\text{cork}}g \implies F_{\text{buoyant}} = m_{\text{cork}}g \implies \rho_{\text{water}}V_{\text{cork submerged}}g = mg = \rho_{\text{cork}}V_{\text{cork}}$$
$$\implies \frac{V_{\text{cork submerged}}}{V_{\text{cork}}} = \frac{\rho_{\text{cork}}}{\rho_{\text{water}}} = \frac{120 \text{ kg/m}^3}{1.00 \times 10^3 \text{ kg/m}^3} = 0.120 = 12\%.$$

c) Consider the cork and aluminum as the system.

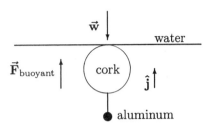

The forces on the system are the same as in parts a) and b) except $\vec{w}$ now includes the weight of the aluminum. When the system is floating and in equilibrium, the total force on it must be zero. Therefore, taking $\hat{j}$ pointing up, we have

$$0\,\text{N} = F_{y\,\text{total}} = F_{\text{buoyant}} - w = F_{\text{buoyant}} - m_{\text{cork}} + m_{\text{Al}})g \implies F_{\text{buoyant}} = m_{\text{cork}} + m_{\text{Al}})g$$

$$\implies \rho_{\text{water}}(V_{\text{cork}} + V_{\text{Al}})g = m_{\text{cork}} + m_{\text{Al}})g \implies \rho_{\text{water}}\left(V_{\text{cork}} + \frac{m_{\text{Al}}}{\rho_{\text{Al}}}\right) = (\rho_{\text{cork}}V_{\text{cork}} + m_{\text{Al}}).$$

Solve the last equation for m_{Al}. After some algebra, you will find

$$m_{Al} = \frac{\rho_{\text{Al}}(\rho_{\text{water}} - \rho_{\text{cork}})}{\rho_{\text{Al}} - \rho_{\text{water}}}V_{\text{cork}}.$$

Substitute the numerical values into this last formula to find

$$m_{Al} = \frac{(2.70 \times 10^3\,\text{kg/m}^3)(1.00 \times 10^3\,\text{kg/m}^3 - 120\,\text{kg/m}^3)}{2.70 \times 10^3\,\text{kg/m}^3 - 1.00 \times 10^3\,\text{kg/m}^3}(3.00 \times 10^{-5}\,\text{m}^3) = 4.2 \times 10^{-2}\,\text{kg}.$$

d) Consider just the cork to be the system. The forces on the cork are:

1. its weight $\vec{w}$, pointing down;

2. the tension $\vec{T}$ of the string, pointing down; and

3. the buoyant force $\vec{F}_{\text{buoyant}}$, pointing up.

The cork is floating in such a way that it is totally submerged and in equilibrium. The total force on it must be zero, so

$$0\,\text{N} = F_{\text{total}} = F_{\text{buoyant}} - W - T = F_{\text{buoyant}} - m_{\text{cork}}g - T$$
$$\implies T = F_{\text{buoyant}} - m_{\text{cork}}g = \rho_{\text{water}}V_{\text{cork}}g - \rho_{\text{cork}}V_{\text{cork}}g = (\rho_{\text{water}} - \rho_{\text{cork}})V_{\text{cork}}g$$
$$= (1.00 \times 10^3\,\text{kg/m}^3 - 120\,\text{kg/m}^3)(3.00 \times 10^{-5}\,\text{m}^3)(9.81\,\text{m/s}^2) = 0.26\,\text{N}.$$

11.45 The forces acting on the band are:

1. its weight $\vec{w}$, pointing down; and

2. the buoyant force $\vec{F}_{\text{buoyant}}$, pointing up.

Let $\hat{j}$ point up. The band is floating and in equilibrium, so the total force on it must be zero. Therefore,

$$0\,\text{N} = F_{y\,\text{total}} = F_{\text{buoyant}} - w = F_{\text{buoyant}} - m_{\text{band}}g \implies F_{\text{buoyant}} = m_{\text{band}}g$$
$$\implies \rho_{\text{mercury}}V_{\text{band submerged}}g = m_{\text{band}}g = \rho_{\text{band}}V_{\text{band}}g \implies \rho_{\text{mercury}}(0.20V_{\text{band}}) = \rho_{\text{band}}V_{\text{band}}$$
$$\implies \rho_{\text{band}} = 0.20\rho_{\text{mercury}} = 0.20(13.6 \times 10^3\,\text{kg/m}^3) = 2.7 \times 10^3\,\text{kg/m}^3.$$

11.49   Let the system be the sand and the cup (of negligible mass). Let $\hat{\mathbf{j}}$ point up. The forces acting on the system are:

1. the force $\vec{\mathbf{F}}_{\text{buoyant}}$ pointing up; and

2. the weight $\vec{\mathbf{w}}$ of the sand pointing down.

The system is in equilibrium, so the total force on it must be zero. Therefore,

$$0\text{ N} = F_{y\,\text{total}} = F_{\text{buoyant}} - w = F_{\text{buoyant}} - m_{\text{sand}}g$$

$$\implies F_{\text{buoyant}} = m_{\text{sand}}g \implies \rho_{\text{water}}V_{\text{cup submerged}}g = m_{\text{sand}}g \implies m_{\text{sand}} = \rho_{\text{water}}V_{\text{cup submerged}}.$$

Let R be the radius of the circular cross section of the cup at the water line and ℓ the depth to which the cone sinks. Since the diameter of the cup is equal to its height, similar triangles imply that the diameter at the water line is equal to ℓ, so $R = \dfrac{\ell}{2}$. Then the volume of the submerged portion of the cup is

$$V_{\text{cup submerged}} = \frac{\pi R^2 \ell}{3} = \frac{\pi \left(\dfrac{\ell}{2}\right)^2 \ell}{3} = \frac{\pi \ell^3}{12}.$$

Hence

$$m_{\text{sand}} = \rho_{\text{water}} \frac{\pi \ell^3}{12} \implies \ell = \sqrt[3]{\frac{12 m_{\text{sand}}}{\pi \rho_{\text{water}}}} = \sqrt[3]{\frac{12(0.040\text{ kg})}{\pi(1.00 \times 10^3 \text{ kg/m}^3)}} = 5.3 \times 10^{-2}\text{ m} = 5.3\text{ cm}.$$

11.53   Use equation 11:20 and assume a contact angle of $0°$:

$$h = \frac{2\gamma \cos \theta_{\text{contact}}}{\rho g r} = \frac{2\gamma \cos 0}{\rho g r} = \frac{2\gamma}{\rho g r} \implies \gamma = \frac{h \rho g r}{2}.$$

For a straw whose inside diameter is 2.0 mm, $r = 1.0 \times 10^{-3}$ m. The height of the water in the straw is about 1.5 cm, so

$$\gamma = \frac{(1.5 \times 10^{-2}\text{ m})(1.00 \times 10^3 \text{ kg/m}^3)(9.81 \text{ m/s}^2)(1.0 \times 10^{-3}\text{ m})}{2} = 0.074\text{ N/m}.$$

11.57   Apply Bernoulli's principle. Let position 1 be outside the building, where the wind speed is v_1, and let position 2 be inside the building, where the wind speed is $v_2 = 0$ m/s. The inside and outside have the same elevation, so $y_1 = y_2$. Thus,

Outside   Inside

$$P_1 + \rho \frac{v_1^2}{2} + \rho g y_1 = P_2 + \rho \frac{v_2^2}{2} + \rho g y_2$$

$$\implies P_1 + \rho \frac{v_1^2}{2} + \rho g y_1 = P_2 + \rho \frac{(0 \text{ m/s})^2}{2} + \rho g y_1$$

$$\implies P_1 + \rho \frac{v_1^2}{2} = P_2.$$

The pressure difference is

$$\Delta P = P_2 - P_1 = \rho \frac{v_1^2}{2} \implies v_1 = \sqrt{\frac{2\Delta P}{\rho}}.$$

The density ρ of air is about 1.3 kg/m^3. Since the inside air pressure is atmospheric pressure, and the pressure difference is 10%, we have

$$\Delta P = 0.10(1.01 \times 10^5 \text{ Pa}) = 1.01 \times 10^4 \text{ Pa} \implies v_1 = \sqrt{\frac{2(1.01 \times 10^4 \text{ Pa})}{1.3 \text{ kg/m}^3}} = 1.2 \times 10^2 \text{ m/s}.$$

11.61   First, use the equation of flow continuity to relate the speeds of the beer in the two pipes. Let d_1 and d_2 be the respective diameters of the two pipes. Then

$$A_1 v_1 = A_2 v_2 \implies \pi \left(\frac{d_1}{2}\right)^2 v_1 = \pi \left(\frac{d_2}{2}\right)^2 v_2 \implies v_2 = \left(\frac{d_1}{d_2}\right)^2 v_1 = \left(\frac{5.0 \times 10^{-2} \text{ m}}{2.0 \times 10^{-2} \text{ m}}\right)^2 v_1 = 6.3 v_1.$$

Now apply Bernoulli's principle, noting that both pipes are at the same elevation, so $y_1 = y_2$:

$$\begin{array}{cc} \text{Larger pipe} & \text{Smaller pipe} \end{array}$$

$$P_1 + \rho \frac{v_1^2}{2} + \rho g y_1 = P_2 + \rho \frac{v_2^2}{2} + \rho g y_2$$

$$\implies P_1 + \rho \frac{v_1^2}{2} = P_2 + \rho \frac{v_2^2}{2}$$

$$\implies 9.60 \times 10^5 \text{ Pa} + \rho \frac{v_1^2}{2} = 5.50 \times 10^5 \text{ Pa} + \rho \frac{(6.3 v_1)^2}{2}.$$

So

$$\rho \frac{40 v_1^2 - v_1^2}{2} = 4.1 \times 10^5 \text{ Pa} \implies (1.00 \times 10^3 \text{ kg/m}^3) \frac{40 v_1^2 - v_1^2}{2} = 4.1 \times 10^5 \text{ Pa}.$$

Solve this last equation for v_1 and find

$$v_1 = 4.6 \text{ m/s}.$$

The flow rate then is

$$A_1 v_1 = \pi \left(\frac{d_1}{2}\right)^2 v_1 = \pi \left(\frac{5.0 \times 10^{-2} \text{ m}}{2}\right)^2 (4.6 \text{ m/s}) = 9.0 \times 10^{-3} \text{ m}^3/\text{s}.$$

11.65

a)   Choose a coordinate system with origin at the base of the fountain (the exit nozzle) and with $\hat{j}$ pointing up. Apply the kinematic equations for motion with a constant acceleration to the water coming out of the fountain. The velocity component at any instant is

$$v_y(t) = v_{y0} + a_y t = v_{y0} - gt.$$

When the water is at greatest height, the velocity component is zero, so

$$0 \text{ m/s} = v_{y0} - gt \implies t = \frac{v_{y0}}{g}.$$

The position at any instant is

$$y(t) = y_0 + v_{y0} t + a_y \frac{t^2}{2} = 0 \text{ m} + v_{y0} t - g \frac{t^2}{2},$$

so, when $y = 20.0 \text{ m}$,

$$20.0 \text{ m} = v_{y0} \left(\frac{v_{y0}}{g}\right) - g \frac{\left(\frac{v_{y0}}{g}\right)^2}{2} = \frac{v_{y0}^2}{2g} \implies v_{y0} = \sqrt{2g(20.0 \text{ m})} = \sqrt{2(9.81 \text{ m/s}^2)(20.0 \text{ m})} = 19.8 \text{ m/s}.$$

b)   The flow rate is the product of the speed and the cross sectional area of the exit nozzle, so

$$\text{Flow rate} = Av = \pi \left(\frac{d}{2}\right)^2 v = \pi \left(\frac{10.0 \times 10^{-2} \text{ m}}{2}\right)^2 (19.8 \text{ m/s}) = 0.156 \text{ m}^3/\text{s} = 156 \text{ liters per second}.$$

c) Apply Bernoulli's principle. Take the top of the reservoir as position 1 and the exit opening as position 2. Use the coordinate system from part a). Both the top of the reservoir and the exit opening are at atmospheric pressure. The speed of the fluid at the top of the reservoir is essentially zero. So

$$\text{top of reservoir} \qquad \text{exit nozzle}$$

$$P_1 + \rho\frac{v_1^2}{2} + \rho g y_1 = P_2 + \rho\frac{v_2^2}{2} + \rho g y_2$$

$$\implies P_1 + \rho\frac{(0 \text{ m/s})^2}{2} + \rho g y_1 = P_1 + \rho\frac{(19.8 \text{ m/s})^2}{2} + \rho g(0 \text{ m})$$

$$\implies \rho g y_1 = \rho\frac{(19.8 \text{ m/s})^2}{2}$$

$$\implies y_1 = \frac{(19.8 \text{ m/s})^2}{2} = 20.0 \text{ m}.$$

11.69 Choose a coordinate system with $\hat{\mathbf{j}}$ pointing up and origin at the surface of the perfume in the dispenser. The difference in pressure between the inside of the bottle and the top of the column of liquid is found from the equation $P(y) = P_0 - \rho g y$. P_0 is atmospheric pressure, so

$$P_0 - P = \rho g y = (900 \text{ kg/m}^3)(9.81 \text{ m/s}^2)(0.050 \text{ m}) = 4.4 \times 10^2 \text{ Pa}.$$

Apply Bernoulli's principle to the air in the chamber (position 1), which is at rest, and to the air above the top of the perfume column (position 2).

$$\text{in the chamber} \qquad \text{top of column}$$

$$P_1 + \rho_{\text{air}}\frac{v_1^2}{2} + \rho_{\text{air}}g y_1 = P_2 + \rho_{\text{air}}\frac{v_2^2}{2} + \rho_{\text{air}}g y_2$$

$$\implies P_1 + \rho_{\text{air}}\frac{(0 \text{ m/s})^2}{2} + \rho_{\text{air}}g(0 \text{ m}) = P_2 + \rho_{\text{air}}\frac{v_2^2}{2} + \rho_{\text{air}}g(0.050 \text{ m})$$

$$\implies P_1 = P_2 + \rho_{\text{air}}\frac{v_2^2}{2} + \rho_{\text{air}}g(0.050 \text{ m}).$$

So

$$P_1 - P_2 = \rho_{\text{air}}\frac{v_2^2}{2} + \rho_{\text{air}}g(0.050 \text{ m}) \implies 4.4 \times 10^2 \text{ Pa} = (1.3 \text{ kg/m}^3)\frac{v_2^2}{2} + (1.3 \text{ kg/m}^3)(9.81 \text{ m/s}^2)(0.050 \text{ m})$$

$$= (1.3 \text{ kg/m}^3)\frac{v_2^2}{2} + 0.64 \text{ Pa} \implies v_2 = 26 \text{ m/s}.$$

Chapter 12

Waves

12.1 The light reaches you at essentially the same instant the cherry bomb explodes. However, it takes 0.50 s for the sound from the explosion to reach you. Therefore, the distance to the explosion is the product of the speed of sound and the length of the time interval:

$$d = vt = (343 \text{ m/s})(0.50 \text{ s}) = 1.7 \times 10^2 \text{ m}.$$

12.5

a) Let t be the length of time that the faster p-waves take to travel from the focus to the point of detection. Then the time that the slower s-waves take to travel the same distance is $t + 240$ s. Since distance is equal to the product of speed and time, we have

$$d = (5.0 \text{ km/s})t = (3.0 \text{ km/s})(t + 240 \text{ s}) \implies t = 360 \text{ s}.$$

Hence the distance to the focus is

$$d = (5.0 \text{ km/s})t = (5.0 \text{ km/s})(360 \text{ s}) = 1.8 \times 10^3 \text{ km}.$$

b) The seismic waves could have come from anywhere within that part of the Earth on the surface of a sphere of radius 1.8×10^3 km centered on the detector.

c) By using seismometers at *two* different locations the location of the focus can be narrowed down to the intersection of *two* spherical surfaces. The intersection of two spherical shells is a circle, and the portion of this circle that lies below the Earth's surface (ignoring the curvature of the Earth) is a semicircle. The projection of this semicircle on the Earth's surface is a line segment. With seismometers located at three different stations, each determining the distance to the quake, the focus of the quake would be located at the intersection of the three semicircles that each pair of stations determine. This intersection, ignoring errors in the model and in measurement, should be a single point.

12.9

a) The period T of the wave is the length of time after which the wave starts to repeat itself. The period of the graph in Figure P:9 is three.

b) The frequency ν of the wave is the reciprocal of the period T, so

$$\nu = \frac{1}{T} = \frac{1}{3.0 \text{ s}} = 0.33 \text{ Hz}.$$

c) Notice that at $x = 0$ m, and as t increases from 0 s, the particle moves toward a value of 2.0 m for Ψ then decreases back to zero, and does all this during 1.0 s. Since the wave is moving at a speed of 3.00 m/s, this behavior must be displayed over a 3.0 m distance in space. The oscillation then remains at zero for 1.0 s corresponding to 3.0 m in space. The oscillation then decreases to -2.0 m, passes through 0 m during the next second corresponding to another 3.0 m distance. The oscillation then repeats itself. The wave is moving toward increasing values of x. Therefore, the waveform must display this pattern as you move toward *decreasing* values of x from the origin. The wavelength λ of the wave is 9.0 m.

d)   The product of the frequency and wave length is

$$\nu\lambda = (0.33 \text{ Hz})(9.0 \text{ m}) = 3.0 \text{ m/s},$$

which is, indeed, the speed of the wave.

12.13

a)   Substitute the given numerical values into Ψ and make the graphs using suitable software or a graphing calculator. The graphs are shown in the following two figures:

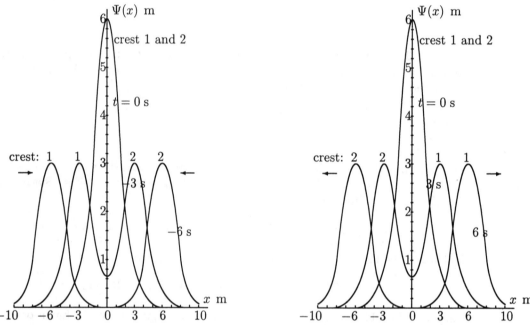

The figure on the left shows Ψ at times $t = -6.00$ s, $t = -3.00$ s, and $t = 0.00$ s. The figure on the right shows Ψ at times $t = 0.00$ s, $t = 3.00$ s, and $t = 6.00$ s. Wave 1 is moving from left to right, wave 2 from right to left.

b)   The constant A is the height (3 m) of each of the two wave crests that approach and then pass through the origin from opposite directions.

c)   The two wave crests overlap at the origin when $t = 0$ s. At this time and place, a larger disturbance is created than that by either crest individually.

d)   Substitute the given numerical values into Ψ and remake the graphs. Notice that increasing the value of α narrows the width of the crests, while increasing v means that the crests travel through a greater distance than before during each 3.00 s time interval.

12.17

a)   The amplitude A of the wave is the coefficient of the cosine term, so $A = 6.00 \times 10^{-3}$ Pa.

b)   The angular wavenumber k is the coefficient of x in the argument of the cosine, so $k = 8.06$ rad/m.

c)   The angular wavenumber *is* the number of wavelengths that fit into 2π m, so 8.06 wavelengths fit into that distance.

d)   The wavelength is

$$\lambda = \frac{2\pi \text{ rad}}{k} = \frac{2\pi \text{ rad}}{8.06 \text{ rad/m}} = 0.780 \text{ m}.$$

e) The angular frequency ω of the wave is the coefficient of t in the argument of the cosine, so $\omega = 2.76 \times 10^3$ rad/s.

f) The frequency is

$$\nu = \frac{\omega}{2\pi} = \frac{2.76 \times 10^3 \text{ rad/s}}{2\pi} = 439 \text{ Hz}.$$

g) The period of the wave is

$$T = \frac{1}{\nu} = \frac{1}{439 \text{ Hz}} = 2.28 \times 10^{-3} \text{ s}.$$

h) The speed of the wave is

$$v = \nu\lambda = (439 \text{ Hz})(0.780 \text{ m}) = 342 \text{ m/s}.$$

i) To find the waveform when $t = 1.00$ s, substitute this value for t in the wavefunction:

$$\Psi(x, 100 \text{ s}) = (6.00 \times 10^{-3} \text{ Pa}) \cos\left[(8.06 \text{ rad/m})x - (2.76 \times 10^3 \text{ rad/s})(1.00 \text{ s})\right]$$
$$= (6.00 \times 10^{-3} \text{ Pa}) \cos\left[(8.06 \text{ rad/m})x - 2.76 \times 10^3 \text{ rad}\right].$$

Since the wavelength is 0.780 m, plot this function using suitable software or graphing calculator from, say, $x = 0$ m to $x = 2$ m.

j) To find the oscillation where $x = 2.00$ m, substitute this value for x in the wavefunction:

$$\Psi(2.00 \text{ m}, t) = (6.00 \times 10^{-3} \text{ Pa}) \cos\left[(8.06 \text{ rad/m})(2.00 \text{ m}) - (2.76 \times 10^3 \text{ rad/s})t\right]$$
$$= (6.00 \times 10^{-3} \text{ Pa}) \cos\left[16.1 \text{ rad} - (2.76 \times 10^3 \text{ rad/s})t\right].$$

Since the period is 2.28×10^{-3} s, plot this function from, say, $t = 0$ s to 6×10^{-3} s.

k) To find the wave disturbance Ψ at $x = 1.00$ m when $t = 0$ s, substitute these numerical values into the wavefunction:

$$\Psi = (6.00 \times 10^{-3} \text{ Pa}) \cos(8.06 \text{ rad}) = -1.23 \times 10^{-3} \text{ Pa}.$$

The negative result implies a rarefaction with an absolute pressure below the average value in the medium in which the sound is propagating.

12.21

a) The wavelength λ is the distance between the ripple crests, so $\lambda = 0.200$ m. The speed of the wave is

$$v = \lambda\nu \implies \nu = \frac{v}{\lambda} = \frac{0.600 \text{ m/s}}{0.200 \text{ m}} = 3.00 \text{ Hz}.$$

b) The amplitude A of the wave is half the peak to trough distance, $A = 0.50 \times 10^{-2}$ m. The angular wavenumber is

$$k = \frac{2\pi \text{ rad}}{\lambda} = \frac{2\pi \text{ rad}}{0.200 \text{ m}} = 31.4 \text{ rad/m}.$$

The angular frequency is

$$\omega = (2\pi \text{ rad})\nu = (2\pi \text{ rad})(3.00 \text{ Hz}) = 18.8 \text{ rad/s}.$$

The wavefunction then is

$$\Psi(x, t) = A\cos(kx - \omega t) = (0.50 \times 10^{-2} \text{ m}) \cos\left[(31.4 \text{ rad/m})x - (18.8 \text{ rad/s})t\right].$$

12.25   The speed v_1 of the waves on the string with tension of magnitude T_1 is $v_1 = \sqrt{\dfrac{T_1}{\mu}}$, so

$$T_1 = \mu v_1^2.$$

If the tension is changed to T_2, then the corresponding wave speed is v_2 and

$$T_2 = \mu v_2^2.$$

Divide T_2 by T_1.

$$\frac{T_2}{T_1} = \frac{\mu v_2^2}{\mu v_1^2} = \frac{v_2^2}{v_1^2} \implies T_2 = \frac{v_2^2}{v_1^2} T_1 = \frac{(45 \text{ m/s})^2}{(35 \text{ m/s})^2} 50 \text{ N} = 83 \text{ N}.$$

12.29   The mass per unit length of the rope is

$$\mu = \frac{m}{\ell} = \frac{0.156 \text{ kg}}{10.00 \text{ m}} = 1.56 \times 10^{-2} \text{ kg/m}.$$

The speed of the waves on the rope is

$$v = \sqrt{\frac{T}{\mu}} = \sqrt{\frac{40.0 \text{ N}}{1.56 \times 10^{-2} \text{ kg/m}}} = 50.6 \text{ m/s}.$$

Since you initiate a pulse first, your pulse travels along the rope a greater distance. The pulses meet 1.00 m from the center of the rope, so your pulse travels 6.00 m, while the later pulse from your buddy travels only 4.00 m. Let t be the time the pulse from your buddy travels the 4.00 m. Then, since the speed of both pulses is 50.6 m/s, the equation for the pulse from your buddy is

$$(50.6 \text{ m/s})t = 4.00 \text{ m}.$$

Your pulse travels along the rope for a time $t + \Delta t$ and travels 6.00 m, so

$$6.00 \text{ m} = (50.6 \text{ m/s})(t + \Delta t) = (50.6 \text{ m/s})t + (50.6 \text{ m/s})\Delta t = 4.00 \text{ m} + (50.6 \text{ m/s})\Delta t$$

$$\implies \Delta t = \frac{6.00 \text{ m} - 4.00 \text{ m}}{50.6 \text{ m/s}} = 3.95 \times 10^{-2} \text{ s}.$$

12.33   The reflected pulses are 4.0 m apart, so the time between them is

$$\Delta t = \frac{4.0 \text{ m}}{343 \text{ m/s}} = 1.2 \times 10^{-2} \text{ s}.$$

The frequency of the pulse is the inverse of this time.

$$\nu = \frac{343 \text{ m/s}}{4.0 \text{ m}} = 86 \text{ Hz}.$$

12.37   The power of the lights is spread uniformly over the lateral surface area of a cylinder of radius r and length ℓ centered on the lamps. The intensity is the power per unit area, so

$$I = \frac{P}{2\pi r \ell}.$$

12.41

a)   The sound level is

$$\beta = (10 \text{ dB}) \log \frac{I}{I_0},$$

where $I_0 = 10^{-12}$ W/m^2 is the threshold of hearing. Use the given information to find the intensity of the source:

$$120 \text{ dB} = (10 \text{ dB}) \log \frac{I}{I_0} \implies 12.0 = \log \frac{I}{I_0} \implies \frac{I}{I_0} = 10^{12.0}$$

$$\implies I = (10^{-12} \text{ W/m}^2)10^{12.0} = 1.00 \text{ W/m}^2.$$

The intensity is related to the power P of the source and the distance r from it by

$$I = \frac{P}{4\pi r^2} \implies P = 4\pi r^2 I = 4\pi (5.0 \text{ m})^2 (1.00 \text{ W/m}^2) = 3.1 \times 10^2 \text{ W}.$$

b)   The sound level is

$$\beta = (10 \text{ dB}) \log \frac{I}{I_0} \implies 90 \text{ dB} = (10 \text{ dB}) \log \frac{I}{I_0} \implies$$

$$9.0 = \log \frac{I}{I_0} \implies \frac{I}{I_0} = 10^{9.0} \implies I = (10^{-12} \text{ W/m}^2)10^{9.0} = 1.0 \times 10^{-3} \text{ W/m}^2.$$

The intensity is related to the power P of the source and the distance r from it by

$$I = \frac{P}{4\pi r^2} \implies r = \sqrt{\frac{P}{4\pi I}} = \sqrt{\frac{3.1 \times 10^2 \text{ W}}{(4\pi)1.0 \times 10^{-3} \text{ W/m}^2}} = 1.6 \times 10^2 \text{ m}.$$

12.45   The sound level is

$$\beta = (10 \text{ dB}) \log \frac{I}{I_0}$$

where $I_0 = 10^{-12}$ W/m^2 is the threshold of hearing. Use the given information to find the intensity of the source:

$$120 \text{ dB} = (10 \text{ dB}) \log \frac{I}{I_0} \implies 12.0 = \log \frac{I}{I_0} \implies \frac{I}{I_0} = 10^{12.0} \implies I = (10^{12} \text{ W/m}^2)10^{-12.0} = 1.00 \text{ W/m}^2.$$

The intensity is related to the power P of the source and the distance r from it by

$$I = \frac{P}{4\pi r^2} \implies P = 4\pi r^2 I = 4\pi (6.0 \text{ m})^2 (1.00 \text{ W/m}^2) = 4.5 \times 10^2 \text{ W}.$$

Now use the power to find the intensity I' at a distance 30.0 m from the source:

$$I' = \frac{P}{4\pi r^2} = \frac{4.5 \times 10^2 \text{ W}}{4\pi (30.0 \text{ m})^2} = 4.0 \times 10^{-2} \text{ W/m}^2.$$

The sound level at 30.0 m is

$$\beta' = (10 \text{ dB}) \log \frac{I'}{I_0} = (10 \text{ dB}) \log \left(\frac{4.0 \times 10^{-2} \text{ W/m}^2}{10^{-12} \text{ W/m}^2} \right)$$

$$= (10 \text{ dB}) \log(4.0 \times 10^{10}) = (10 \text{ dB}) \log(4.0) + (10 \text{ dB})10 = 6.0 \text{ dB} + 100 \text{ dB} = 106 \text{ dB}.$$

12.49

a)   Since you run *from* the room, the frequency heard by your classmates is *less* than the frequency you emit. They hear

$$2.200 \times 10^3 \text{ Hz} - 20 \text{ Hz} = 2.180 \times 10^3 \text{ Hz}.$$

b)   The observers are at rest so $v_{\text{obs}} = 0 \text{ m/s}$, and the observed frequency is less than the source frequency. Hence, the equation for the acoustic Doppler effect is

$$\nu' = \nu \frac{v}{v + v_{\text{source}}} \implies 2.180 \times 10^3 \text{ Hz} = (2.200 \times 10^3 \text{ Hz}) \frac{343 \text{ m/s}}{343 \text{ m/s} + v_{\text{source}}} \implies v_{\text{source}} = 3 \text{ m/s}.$$

12.53

a)   The driver of the ambulance is not moving with respect to the siren, so the frequency heard is the "true" frequency, 1200 Hz.

b)   Convert the speeds from km/h to m/s:

$$v_{\text{source}} = 130.0 \text{ km/h} = (130.0 \text{ km/h}) \left(\frac{10^3 \text{ m}}{\text{km}} \right) \left(\frac{\text{h}}{3600 \text{ s}} \right) = 36.11 \text{ m/s},$$

$$v_{\text{obs}} = 140.0 \text{ km/h} = (140.0 \text{ km/h}) \left(\frac{10^3 \text{ m}}{\text{km}} \right) \left(\frac{\text{h}}{3600 \text{ s}} \right) = 38.89 \text{ m/s}.$$

Use the general equation for the Doppler effect for a moving source and observer,

$$\nu' = \nu \frac{v \pm v_{\text{obs}}}{v \mp v_{\text{source}}}.$$

Think of the observer (the lawyer) as being at rest. The source is moving away from the observer, so the frequency heard by the observer is less than the "true" frequency, and therefore the $+$ sign is used in the denominator . Similarly, think of the source as being at rest, then the observer is approaching the source, a higher frequency is heard, so the $+$ sign is used in the numerator. Therefore

$$\nu' = (1200 \text{ Hz}) \left(\frac{343 \text{ m/s} + 38.89 \text{ m/s}}{343 \text{ m/s} + 36.11 \text{ m/s}} \right) = 1.21 \times 10^3 \text{ Hz}.$$

c)   Use the general equation for the Doppler effect for a moving source and observer,

$$\nu' = \nu \frac{v \pm v_{\text{obs}}}{v \mp v_{\text{source}}}.$$

Think of the observer (the lawyer) as being at rest. The source is moving toward the observer, so the frequency heard by the observer is higher than the "true" frequency, and so the $-$ sign is used in the denominator. Similarly, think of the source as being at rest, then the observer is receding from the source, a lower frequency is heard, so the $-$ sign is used in the numerator. Therefore

$$\nu' = (1200 \text{ Hz}) \left(\frac{343 \text{ m/s} - 38.89 \text{ m/s}}{343 \text{ m/s} - 36.11 \text{ m/s}} \right) = 1.19 \times 10^3 \text{ Hz}.$$

12.57   Use the result of part b) of problem 56:

$$\frac{\nu'}{\nu} = 1 + 2 \frac{v_{\text{target}}}{v} \implies \nu' = (30.0 \text{ kHz}) \left(1 + 2 \frac{1.00 \text{ m/s}}{343 \text{ m/s}} \right) = 30.2 \text{ kHz}.$$

12.61

a) Convert the speed of the airplane from km/h to m/s:

$$2150\text{ km/h} = (2150\text{ km/h})\left(\frac{10^3\text{ m}}{\text{km}}\right)\left(\frac{\text{h}}{3600\text{ s}}\right) = 597.2\text{ m/s}.$$

The Mach number of the airplane is

$$\text{Mach number} = \frac{v_{\text{source}}}{v} = \frac{597.2\text{ m/s}}{2.9\times10^2\text{ m/s}} = 2.1.$$

b) The half-angle ϕ of the Mach cone satisfies

$$\sin\phi = \frac{v}{v_{\text{source}}} = \frac{2.9\times10^2\text{ m/s}}{597.2\text{ m/s}} = 0.49 \implies \phi = 29°.$$

12.65 The nodes of a standing wave are separated by half a wavelength:

$$\ell = \frac{\lambda}{2} \implies \lambda = 2\ell.$$

The speed of sound is

$$v = \nu\lambda = \nu(2\ell) = 2\ell\nu.$$

12.69

a) The first wavefunction is propagating toward increasing values of x, while the second is propagating toward decreasing values of x. The two waves have the same amplitude, frequency, wavelength, and speed.

b) The two waves will form a standing wave pattern. The distance between the nodes of the wave is one-half the wavelength. The angular wavenumber k is 9.00 rad/s so

$$k = \frac{2\pi}{\lambda} \implies \lambda = \frac{2\pi\text{ rad/s}}{k} = \frac{2\pi\text{ rad/s}}{9.00\text{ rad/m}} = 0.698\text{ m}.$$

Hence, the distance between the nodes is

$$\frac{0.698\text{ m}}{2} = 0.349\text{ m}.$$

c) The standing wave has the form

$$\Psi(x,t) = 2A\cos kx\cos\omega t.$$

The locations of the first three nodes along the positive x-axis are where $\cos kx$ is zero, i.e., where

(1) $$kx = \frac{\pi}{2}\text{ rad} \implies x = \frac{\pi}{2k}\text{ rad} = \frac{\pi}{2(9.00\text{ rad/m})} = 0.175\text{ m},$$

(2) $$kx = \frac{3\pi}{2}\text{ rad} \implies x = \frac{3\pi}{2k}\text{ rad} = \frac{3\pi}{2(9.00\text{ rad/m})} = 0.524\text{ m},$$

and

(3) $$kx = \frac{5\pi}{2}\text{ rad} \implies x = \frac{5\pi}{2k}\text{ rad} = \frac{5\pi\text{ rad}}{2(9.00\text{ rad/m})} = 0.873\text{ m}.$$

12.73   The wavelength λ of the fundamental vibration of a string of length ℓ is twice the length of the string,

$$\lambda = 2\ell.$$

The speed of the waves on the string is

$$v = \sqrt{\frac{T}{\mu}}.$$

The speed is also the product of the frequency and wavelength,

$$v = \nu\lambda \implies \nu = \frac{1}{\lambda}v = \frac{1}{2\ell}\sqrt{\frac{T}{\mu}}.$$

If the string is shortened to a length $\alpha\ell$, then the corresponding fundamental frequency is reduced to

$$\nu' = \frac{1}{2\alpha\ell}\sqrt{\frac{T}{\mu}}.$$

Therefore, dividing the expression for ν' by the expression for ν,

$$\frac{\nu'}{\nu} = \frac{1}{\alpha} \implies \nu' = \frac{\nu}{\alpha}.$$

12.77

a)   and

b)   Graph the sum of the first few (e.g., six) terms of the Fourier series by hand (ugh!), using software, or using a graphing calculator. The more terms you include the closer the graph approaches the following ideal limiting *square wave* shape.

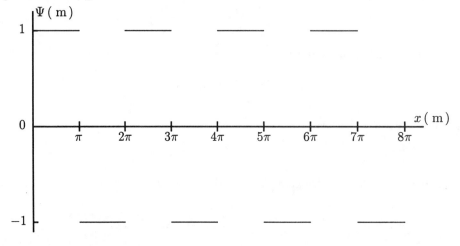

c)   The wavelength of the periodic waveform is the distance along the x-axis between repeating points of the waveform. It is the same as the "period", except now its dimension is length rather than time, since Ψ is a function of distance rather than time. In this case the wavelength is 2π m ≈ 6.3 m.

Chapter 13

The First Law of Thermodynamics

13.1   Use equation 13.5 from the text:

$$t_{\text{celcius}} = \frac{5\,°\text{C}}{9\,°\text{F}}(t_{\text{farenheit}} - 32\,°\text{F}) = \frac{5\,°\text{C}}{9\,°\text{F}}(98.6\,°\text{F} - 32\,°\text{F}) = 37.0\,°\text{C}.$$

13.5   To convert from °C to °F, multiply by $\dfrac{9\,°\text{F}}{5\,°\text{C}}$ and then add exactly 32 °F to account for the different locations of zero on each scale. If $T = -25\,°\text{C}$, then the corresponding Farenheit temperature is

$$\frac{9\,°\text{F}}{5\,°\text{C}}(-25\,°\text{C}) + 32\,°\text{F} = -13\,°\text{F}.$$

13.9   This problem is weird! It requires us to add together three *numbers*, each representing the temperature according to a *different* scale, in order to end up with 415.15! Its a bit like adding apples, pears, and oranges, but here we go.

The condition is

(1)
$$t_{\text{celsius}} + t_{\text{kelvin}} + t_{\text{fahrenheit}} = 495.15.$$

We won't even try to say what the units on the right hand side are!

The relation between the kelvin and celsius scales may be written as

(2)
$$-t_{\text{celsius}} + t_{\text{kelvin}} = 273.15,$$

and the relation between Celsius and Fahrenheit is

(3)
$$-\frac{9}{5}t_{\text{celsius}} + t_{\text{fahrenheit}} = 32.00.$$

These are three simultaneous equations in the three unknowns t_{celsius}, t_{kelvin}, and $t_{\text{farenheit}}$. One way of solving them is to solve (2) for t_{kelvin} in terms of t_{celsius} and solve (3) for $t_{\text{farenheit}}$ in terms of t_{celsius}, then substitute these expressions into (1), to obtain

$$t_{\text{celsius}} + (t_{\text{celsius}} + 273.15) + \left(\frac{9}{5}t_{\text{celsius}} + 32\right) = 495.15.$$

Now solve this for t_{celsius} and find $t_{\text{celsius}} = 50.00$. Use this result in (2) and (3) to find t_{kelvin} and t_{celsius}. After doing this and putting the units back on, the result is

$$t_{\text{celsius}} = 50.00\,°\text{C}, \quad t_{\text{kelvin}} = 323.15\,\text{K}, \quad \text{and} \quad t_{\text{farenheit}} = 122.00\,°\text{F}.$$

13.13   Determine the change in length from the equation for linear expansion:

$$\Delta\ell = \alpha\ell\Delta T.$$

The coefficient of linear expansion for steel is found from Table 13.1 to be $\alpha = 12 \times 10^{-6}$ per degree Kelvin. So

$$\Delta\ell = (12 \times 10^{-6} \text{ K}^{-1})(100.000 \text{ m})(10 \text{ K}) = 1.2 \times 10^{-2} \text{ m}.$$

The percentage error is

$$\frac{\Delta\ell}{\ell}(100\%) = \left(\frac{1.2 \times 10^{-2} \text{ m}}{100.00 \text{ m}}\right) 100\% = \left(1.2 \times 10^{-4}\right) 100\% = 0.012\%.$$

13.17

a)   The fundamental frequency ν of a wire with a tension of magnitude T, mass per unit length μ, and length ℓ is

(1) $$\nu = \frac{1}{2\ell}\sqrt{\frac{T}{\mu}}.$$

Differentiate ν with respect to ℓ to find the rate of change of ν with respect to ℓ:

$$\frac{d\nu}{d\ell} = -\frac{1}{2\ell^2}\sqrt{\frac{T}{\mu}}.$$

Then, for a small change $\Delta\ell$ in ℓ, the corresponding change $\Delta\nu$ in ν is

(2) $$\Delta\nu \approx \left(-\frac{1}{2\ell^2}\sqrt{\frac{T}{\mu}}\right)\Delta\ell.$$

Divide equation (2) by equation (1):

$$\frac{\Delta\nu}{\nu} \approx -\frac{\Delta\ell}{\ell} \implies \Delta\ell = -\ell\frac{\Delta\nu}{\nu}.$$

The frequency decreases by 0.20 Hz so $\Delta\nu = -0.20$ Hz. Therefore

$$\Delta\ell \approx -(1.000 \text{ m})\left(\frac{-0.20 \text{ Hz}}{261.6 \text{ Hz}}\right) = 7.6 \times 10^{-4} \text{ m}.$$

b)   The change in length is related to the temperature change by

$$\Delta\ell \approx \alpha\ell\Delta T \implies \Delta T \approx \frac{\Delta\ell}{\alpha\ell} = \frac{7.6 \times 10^{-4} \text{ m}}{(12 \times 10^{-6} \text{ K}^{-1})(1.000 \text{ m})} = 63 \text{ K}.$$

13.21   The change in the radius is

$$\Delta r = \alpha r\Delta T \implies \Delta T = \frac{\Delta r}{\alpha r} = \frac{0.005 \text{ m}}{(12 \times 10^{-6} \text{ K}^{-1})(0.995 \text{ m})} = 4 \times 10^2 \text{ K}.$$

13.25

$$\frac{dA}{dr} = \frac{d}{dr}\pi r^2 = 2\pi r \implies \Delta A \approx (2\pi r)\Delta r.$$

If the radius changes because of a change ΔT in temperature, then

$$\Delta r = \alpha r\Delta T \implies \Delta A \approx 2\pi r(\alpha r\Delta T) = 2\alpha\pi r^2\Delta T = 2\alpha A\Delta T.$$

13.29 Young's modulus E is the ratio of the stress to strain, so

$$E = \frac{\dfrac{F}{A}}{\dfrac{|\Delta\ell|}{\ell}} \implies F = EA\frac{|\Delta\ell|}{\ell} = EA\frac{|\alpha\ell\Delta T|}{l} = \alpha EA|\Delta T|$$

$$= (12 \times 10^{-6} \text{ K}^{-1})(20 \times 10^{10} \text{ Pa})(\pi(0.0050 \text{ m})^2)(80 \text{ K}) = 1.5 \times 10^4 \text{ N}.$$

13.33 Begin with the ideal gas law:

$$PV = nRT.$$

The density ρ of the gas is the mass per unit volume,

$$\rho = \frac{m}{V} \implies V = \frac{m}{\rho}.$$

Substitute this expression for V into the ideal gas law and solve for ρ:

$$P\frac{m}{\rho} = nRT \implies \rho = \frac{Pm}{nRT} = \left(\frac{m}{nRT}\right)P.$$

Hence, for constant temperature, the density is proportional to the pressure.

13.37 Start from the ideal gas law $PV = nRT$ and solve for P:

(1) $$P = \frac{nRT}{V}.$$

Let $V = 3.0 \text{ m}^3$. There is one particle within this volume, so the number of moles present is

$$n = \frac{N}{N_A} = \frac{1 \text{ particle}}{6.02 \times 10^{23} \text{ particles/mol}} = 1.66 \times 10^{-24} \text{ mol}.$$

Substitute this value of n into equation (1):

$$P = \frac{(1.66 \times 10^{-24} \text{ mol})(8.315 \text{ J/mol·K})(3.0 \text{ K})}{3.0 \text{ m}^3} = 1.4 \times 10^{-23} \text{ Pa}.$$

13.41 Begin with the ideal gas law, using the initial temperature of the gas at the surface of the lake:

$$P_1 V_1 = nRT_1 \implies nR = \frac{P_1 V_1}{T_1}.$$

Since the amount of gas is the same when the diving bell is submerged,

$$\frac{P_1 V_1}{T_1} = \frac{P_2 V_2}{T_2} \implies P_2 = \frac{T_2 V_1 P_1}{T_1 V_2}.$$

When the volume is half the initial volume $V_2 = \dfrac{V_1}{2}$ so, substituting this for V_2,

$$P_2 = \frac{T_2 V_1 P_1}{T_1 \dfrac{V_1}{2}} = \frac{2T_2 P_1}{T_1} = \frac{2(277 \text{ K})(1.013 \times 10^5 \text{ Pa})}{293 \text{ K}} = 1.92 \times 10^5 \text{ Pa}.$$

Choose a coordinate system with origin at the water's surface and $\hat{\mathbf{j}}$ pointing up. (Note that with this choice, y is negative below the water's surface.) The pressure as a function of elevation within the fluid water is given by

$$P = P_0 - \rho g y. \implies y = \frac{P_0 - P}{\rho g}$$

$$= \frac{1.013 \times 10^5 \text{ Pa} - 1.92 \times 10^5 \text{ Pa}}{(1.00 \times 10^3 \text{ kg/m}^3)(9.81 \text{ m/s}^2)} = \frac{-0.91 \times 10^5 \text{ Pa}}{(1.00 \times 10^3 \text{ kg/m}^3)(9.81 \text{ m/s}^2)} = -9.3 \text{ m}.$$

Thus, the water has risen halfway up the diving bell at the point where the bell is 9.3 m below the surface.

13.45

a) and

b) The total heat transfer to the ice and water is zero, since the combined system is assumed isolated. Assume, provisionally, that all the ice melts. The ice must be warmed from $-10.0\,^{\circ}\mathrm{C}$ to $0\,^{\circ}\mathrm{C}$, then melted to water at the same temperature, and finally, warmed from $0\,^{\circ}\mathrm{C}$ to some final temperature t_f. On the other hand, the water must cool from $20\,^{\circ}\mathrm{C}$ to the final temperature t_f. Since the units for temperature *changes* are the same on both the kelvin and celsius scales, we may use the data from Tables 13.4 and 13.5 directly (on pages 602 and 604 of the text). Thus,

$$0\text{ J} = m_{\mathrm{ice}}c_{\mathrm{ice}}[0\,^{\circ}\mathrm{C} - (-10\,^{\circ}\mathrm{C})] + m_{\mathrm{ice}}L_f + m_{\mathrm{ice}}c_{\mathrm{water}}(t_{\mathrm{celsius}} - 0\,^{\circ}\mathrm{C})$$
$$+ m_{\mathrm{original\ water}}c_{\mathrm{water}}(t_{\mathrm{celsius}} - 20.0\,^{\circ}\mathrm{C})$$
$$= (.500\text{ kg})(2050\text{ J/kg·K})(10.00\text{ K}) + (0.500\text{ kg})(3.335\times10^5\text{ J/kg}) + (0.500\text{ kg})(4186\text{ J/kg·K})t_{\mathrm{celsius}}$$
$$+ (4.00\text{ kg})(4186\text{ J/kg·K})(t_{\mathrm{celsius}} - 20.0\,^{\circ}\mathrm{C})$$
$$= 1.03\times10^4\text{ J} + 1.67\times10^5\text{ J} + (2.09\times10^3\text{ J/K})t_{\mathrm{celsius}} + (1.67\times10^4\text{ J/K})t_{\mathrm{celsius}} - 3.35\times10^5\text{ J}$$
$$\implies (1.88\text{ J/K})t_{\mathrm{celsius}} = 1.58\text{ J} \implies t_{\mathrm{celsius}} = 8.40\,^{\circ}\mathrm{C}.$$

Since the amount of heat transfer from the water in cooling to $8.40\,^{\circ}\mathrm{C}$ is enough to melt all the ice and then raise the temperature of the melted ice to $8.40\,^{\circ}\mathrm{C}$, all the ice melts. (If, in solving for t_{celsius}, we had ended up with a negative answer, then we would conclude that not all the ice melts.)

13.49

a) The mass of water in the tank is

$$m = \rho_{\mathrm{water}}V = (1.00\times10^3\text{ kg/m}^3)(0.50\text{ m}^3) = 5.0\times10^2\text{ kg}.$$

The heat transfer required is

$$Q = mc\Delta T = (5.0\times10^2\text{ kg})(4186\text{ J/kg·K})(45\text{ K}) = 9.4\times10^7\text{ J}.$$

b) The heating element provides energy at the rate of $5.0\times10^3\text{ W} = 5.0\times10^3\,\dfrac{\text{J}}{\text{s}}$. Therefore, the time t required to heat the water satisfies

$$\left(5.0\times10^3\,\frac{\text{J}}{\text{s}}\right)t = 9.4\times10^7\text{ J} \implies t = \frac{9.4\times10^7\text{ J}}{5.0\times10^3\,\frac{\text{J}}{\text{s}}}\left(\frac{\text{h}}{3600\text{ s}}\right) = 5.2\text{ h}.$$

13.53   The total surface area of the mobile home is

$$A = 2(3.0\text{ m})(2.5\text{ m}) + 2(25.0\text{ m})(3.0\text{ m}) + 2(25.0\text{ m})(2.5\text{ m}) = 15\text{ m}^2 + 150\text{ m}^2 + 125\text{ m}^2 = 290\text{ m}^2.$$

The heat flow is

$$\frac{dQ}{dt} = \frac{A\Delta T}{R} = \frac{(290\text{ m}^2)(40\text{ K})}{3.0\text{ m}^2\text{K/W}} = 3.9\times10^3\text{ W}.$$

13.57

a) In the steady state, the heat flow through the two materials is the same, so the ratio is 1.

b) The thermal conductivity k of each material is related to its thickness d and R-value by

$$R_{\mathrm{wood}} = \frac{d_{\mathrm{wood}}}{k_{\mathrm{wood}}}, \qquad R_{\mathrm{cork}} = \frac{d_{\mathrm{cork}}}{k_{\mathrm{cork}}}.$$

The thicknesses d_{wood} and d_{cork} are the same. So

$$\frac{R_{\text{wood}}}{R_{\text{cork}}} = \frac{\dfrac{d_{\text{wood}}}{k_{\text{wood}}}}{\dfrac{d_{\text{cork}}}{k_{\text{cork}}}} = \frac{k_{\text{cork}}}{k_{\text{wood}}} = \frac{1}{3}.$$

Let T be the temperature at the cork-wood interface. The heat flow from the hot reservoir through the wood is

$$\frac{dQ}{dt} = \frac{A(T_H - T)}{R_{\text{wood}}} \qquad \text{text, equation (13.38) on page 609,}$$

and that through the cork to the cold reservoir it is

$$\frac{dQ}{dt} = \frac{A(T - T_C)}{R_{\text{cork}}} \qquad \text{text, equation (13.38).}$$

Since the heat flows are equal in the steady state, we have

$$\frac{A(T_H - T)}{R_{\text{wood}}} = \frac{A(T - T_C)}{R_{\text{cork}}}.$$

Multiply both sides by $\dfrac{R_{\text{wood}}}{A}$ and solve for $T_H - T$ in terms of $T_H - T_C$:

$$T_H - T = \frac{R_{\text{wood}}}{R_{\text{cork}}}(T - T_C) = \frac{1}{3}(T - T_C) = \frac{1}{3}(T - T_H + T_H - T_C) = \frac{1}{3}(T_H - T_C) - \frac{1}{3}(T_H - T)$$

$$\implies \frac{4}{3}(T_H - T) = \frac{1}{3}(T_H - T_C)$$

$$\implies T_H - T = \frac{1}{4}(T_H - T_C)$$

$$\implies 25\,^\circ\text{C} - T = \frac{1}{4}[25\,^\circ\text{C} - (-20\,^\circ\text{C})] = 11\,^\circ\text{C}$$

$$\implies T = 14\,^\circ\text{C}.$$

13.61 The mass of the animal is equal to the product of its average density and its volume, $m = \rho V$. The volume is proportional to the cube of some "characteristic length" ℓ associated with the animal. Thus

$$m \propto \ell^3 \implies \ell \propto m^{\frac{1}{3}}.$$

The surface area A of the animal is proportional to the square of the characteristic length,

$$A \propto \ell^2 \implies A \propto m^{\frac{2}{3}},$$

and the heat flow is proportional to the surface area, so

$$\frac{dQ}{dt} \propto m^{\frac{2}{3}}.$$

The ratio of the heat flow from a 1.10×10^3 kg steer to that from a 60 kg physics student then is

$$\left(\frac{1.20 \times 10^3 \text{ kg}}{60 \text{ kg}} \right)^{\frac{2}{3}} = 7.4.$$

13.65 In each case the work done by the gas is the area under the curve on the P–V diagram. The volumes must be converted to m^3 and the pressures to Pa.

a) The pressure is constant along path (a), so the work is

$$W = P\Delta V = (6.00 \text{ atm}) \left(\frac{1.013 \times 10^5 \text{ Pa}}{\text{atm}} \right) (5.00 \text{ liter}) \left(\frac{\text{m}^3}{10^3 \text{ liter}} \right) = 3.04 \times 10^3 \text{ J}.$$

b) Along the vertical segment of the path, no work is done because there is no change in volume. Along the horizontal segment, the pressure is constant, so the work done by the gas is

$$W = P\Delta V = (3.00 \text{ atm}) \left(\frac{1.013 \times 10^5 \text{ Pa}}{\text{atm}} \right) (5.00 \text{ liter}) \left(\frac{\text{m}^3}{10^3 \text{ liter}} \right) = 1.52 \times 10^3 \text{ J}.$$

c) The area under the curve is the sum of the area under the horizontal part of path (b) plus the triangular area on top of it:

$$W = 1.52 \times 10^3 \text{ J} + \frac{1}{2}(3.00 \text{ atm}) \left(\frac{1.013 \times 10^5 \text{ Pa}}{\text{atm}} \right) (5.00 \text{ liter}) \left(\frac{\text{m}^3}{10^3 \text{ liter}} \right)$$
$$= 1.52 \times 10^3 \text{ J} + 760 \text{ J}$$
$$= 2.28 \times 10^3 \text{ J}.$$

d) This work is the negative of the work in part a), $W = -3.04 \times 10^3$ J.

e) This work is the negative of the work in part b), $W = -1.52 \times 10^3$ J.

f) This work is the negative of the work in part c), $W = -2.28 \times 10^3$ J.

13.69

a) The gas does positive work as it expands from 2.00 liters to 6.00 liters, and this work is equal to the area under this portion of the curve. As the gas is compressed from 6.00 liters back to 2.00 liters along the rest of the path, the gas does negative work equal to the area under this portion of the curve. The negative work is greater in magnitude than the positive work done during the expansion. Thus, the total work done by the gas is the negative of the area of the semicircle. Converting pressure to Pa and volume to m^3:

$$W = -\frac{1}{2}\pi(4.50 \text{ atm}) \left(\frac{1.013 \times 10^5 \text{ Pa}}{\text{atm}} \right) (4.00 \text{ liter}) \left(\frac{\text{m}^3}{10^3 \text{ liter}} \right) = -2.86 \times 10^3 \text{ J}.$$

b) When the gas retraces the curve from state 2 back to state 1, the work done by the gas is the negative of the work calculated in part a), so $W = 2.86 \times 10^3$ J.

13.73

a) Use the CWE theorem to determine the speed of the ball the instant before its first impact. Use a coordinate system with origin on the floor at the point of impact, and with $\hat{\mathbf{j}}$ pointing up. The only force acting on the ball during its descent is the conservative gravitational force, whose work is accounted for by the change in the appropriate gravitational potential energy function. Hence the CWE theorem becomes

$$0 \text{ J} = W_{\text{conservative}} = \Delta\text{KE} + \Delta\text{PE} = \left(m\frac{v^2}{2} - 0 \text{ J} \right) + (0 \text{ J} - mgy_i)$$
$$\Longrightarrow v = \sqrt{2gy_i} = \sqrt{2(9.81 \text{ m/s}^2)(10.0 \text{ m})} = 14.0 \text{ m/s}.$$

The ball rebounds with a speed v' equal to 0.7071 of the impact speed,

$$v' = 0.7071(14.0 \text{ m/s}) = 9.90 \text{ m/s}.$$

Use the CWE theorem again to find the height to which the ball rises after the first impact:

$$0 \text{ J} = \Delta\text{KE} + \Delta\text{PE} = \left(0 \text{ J} - m\frac{v'^2}{2} \right) + (mgy_f - 0 \text{ J}) \Longrightarrow y_f = \frac{v'^2}{2g} = \frac{(9.90)^2}{2(9.81 \text{ m/s}^2)} = 5.00 \text{ m}.$$

b)   The change in the kinetic energy of the ball before and after the impact with the floor is

$$\Delta KE = KE_{after} - KE_{before} = m\frac{v'^2}{2} - m\frac{v^2}{2} = \frac{m}{2}(v'^2 - v^2) = \frac{0.100 \text{ kg}}{2}((9.90 \text{ m/s})^2 - (14.0 \text{ m/s})^2) = 4.9 \text{ J}.$$

The loss of kinetic energy by the ball appears as an increase in its internal energy, *as if* heat transfer occurred to the ball. The temperature rise of the ball is found from

$$Q = mc\Delta T \implies 4.9 \text{ J} = (0.100 \text{ kg})(460 \text{ J/kg·K})\Delta T \implies \Delta T = 0.11 \text{ K}.$$

c)   The total initial mechanical energy of the ball is its initial gravitational potential energy,

$$mgy_i = (0.100 \text{ kg})(9.81 \text{ m/s}^2)(10.00 \text{ m}) = 9.81 \text{ J}.$$

This energy increases the total internal energy of the ball *as if* heat transfer of the same amount occured to the ball. The rise in temperature of the ball is found from

$$Q = mc\Delta T \implies 9.81 \text{ J} = (0.100 \text{ kg})(460 \text{ J/kg·K})\Delta T \implies \Delta T = 0.213 \text{ K}.$$

13.77   The initial total mechanical energy of the water bomb appears as an increase in its internal energy, *as if* heat transfer of the same amount occured to it. Hence

$$mgy_i = mc\Delta T \implies \Delta T = \frac{gy_i}{c} = \frac{(9.81 \text{ m/s}^2)(170 \text{ m})}{4186 \text{ J/kg·K}} = 0.398 \text{ K}.$$

Chapter 14

Kinetic Theory

14.1 The time necessary to complete the count is

$$6.022 \times 10^{23} \text{ s} = (6.022 \times 10^{23} \text{ s}) \left(\frac{\text{h}}{3600 \text{ s}} \right) \left(\frac{\text{d}}{24 \text{ h}} \right) \left(\frac{\text{y}}{365.25 \text{ d}} \right) = 1.908 \times 10^{16} \text{ y} .$$

14.5 The average kinetic energy for one particle is

$$\text{KE}_{\text{ave}} = \frac{3}{2} kT = \frac{3}{2} (1.381 \times 10^{-23} \text{ J/K})(15 \times 10^6 \text{ K}) = 3.1 \times 10^{-16} \text{ J} .$$

14.9 Molecular hydrogen gas has a molar mass of $M_{\text{hydrogen}} = 2 \text{ g/mol} = 2 \times 10^{-3} \text{ kg/mol}$. Its rms speed is

(1)
$$v_{\text{rms hydrogen}} = \sqrt{\frac{3RT}{M_{\text{hydrogen}}}} .$$

So

$$v_{\text{rms hydrogen}} = \sqrt{\frac{3(8.315 \text{ J/mol·K})(300 \text{ K})}{2 \times 10^{-3} \text{ kg/mol}}} = 1.93 \times 10^3 \text{ m/s} .$$

The molar mass of molecular oxygen gas is $M_{\text{oxygen}} = 32 \text{ g/mol} = 32 \times 10^{-3} \text{ kg/mol}$. Its rms speed is

(2)
$$v_{\text{rms oxygen}} = \sqrt{\frac{3RT}{M_{\text{oxygen}}}} .$$

Divide equation (1) by equation(2).

$$\frac{v_{\text{rms hydrogen}}}{v_{\text{oxygen}}} = \sqrt{\frac{M_{\text{oxygen}}}{M_{\text{hydrogen}}}} = \sqrt{\frac{32 \times 10^{-3} \text{ kg/mol}}{2 \times 10^{-3} \text{ kg/mol}}} = 4 .$$

So, when at the same temperature,

$$v_{\text{rms hydrogen}} = 4 v_{\text{rms oxygen}} .$$

The rms speed of hydrogen gas molecules is 4 times that of oxygen gas molecules.

14.13 According to the impulse-momentum theorem of Chapter 9, the impulse given to the molecule by your nose is equal to the change in the momentum of the particle. The impulse given your nose by the molecule is the negative of the impulse given to the particle by your nose (from Newton's third law). Hence the magnitude I of the impulse given by the particle to your nose is $I = \Delta p = 2m|v_{\text{rms}}|$. The rms speed of the particles in the gas is

$$v_{\text{rms}} = \sqrt{\frac{3RT}{M}}.$$

The molar mass of molecular oxygen gas is $M_{\text{oxygen}} = 32$ g/mol $= 32 \times 10^{-3}$ kg/mol. The mass m of an individual oxygen molecule is

$$m = \frac{32 \times 10^{-3} \text{ kg/mol}}{6.022 \times 10^{23} \text{ particle/mol}} = 5.314 \times 10^{-26} \text{ kg}.$$

Substituting in the numerical values, we find

$$v_{\text{rms}} = \sqrt{\frac{3(8.315 \text{ J/mol·K})(300 \text{ K})}{32 \times 10^{-3} \text{ kg/mol}}} = 4.84 \times 10^2 \text{ m/s}.$$

Therefore the magnitude of the impulse given to your nose is

$$I = 2mv_{\text{rms}} = 2(5.314 \times 10^{-26} \text{ kg})(4.84 \times 10^2 \text{ m/s}) = 5.14 \times 10^{-23} \text{ N·s}.$$

You probably won't say "ouch."

14.17

a) Use the ideal gas law to find the volume:

$$PV = nRT \implies V = \frac{nRT}{P} = \frac{(1.00 \text{ mol})(8.315 \text{ J/mol·K})(300 \text{ K})}{1.013 \times 10^5 \text{ Pa}} = 2.46 \times 10^{-2} \text{ m}^3.$$

b) Imagine each particle inside a cube of side ℓ. Then ℓ is also the approximate distance between the particles. Since there is one mole of the gas, we must have

$$N_A \ell^3 \approx V \implies \ell \approx \sqrt[3]{\frac{V}{N_A}} = \sqrt[3]{\frac{2.46 \times 10^{-2} \text{ m}^3}{6.022 \times 10^{23}}} = 3.4 \times 10^{-9} \text{ m}.$$

14.21 The molar mass of nitrogen gas (N_2) is $M = 28$ g/mol $= 28 \times 10^{-3}$ kg/mol. The rms speed of the particles in the gas is

$$v_{\text{rms}} = \sqrt{\frac{3RT}{M}} \implies \frac{dv_{\text{rms}}}{dT} = \frac{1}{2\sqrt{\frac{3RT}{M}}} \frac{3R}{M} = \frac{1}{2}\sqrt{\frac{3R}{MT}}.$$

Therefore, for a small change ΔT in T, the corresponding change Δv_{rms} is given approximately by

$$\Delta v_{\text{rms}} \approx \left(\frac{1}{2}\sqrt{\frac{3R}{MT}}\right)\Delta T.$$

Hence, for $\Delta T = 1.0$ K and the other given numerical values,

$$\Delta v_{\text{rms}} = \left(\frac{1}{2}\sqrt{\frac{3(8.315 \text{ J/mol·K})}{(28 \times 10^{-3} \text{ kg/mol})(300.0 \text{ K})}}\right)(1.0 \text{ K}) = 0.86 \text{ m/s}.$$

14.25 The molar mass of helium is $M = 4$ g/mol $= 4 \times 10^{-3}$ kg/mol. The rms speed of the particles in the gas is

$$v_{\text{rms}} = \sqrt{\frac{3RT}{M}} \implies T = \frac{Mv_{\text{rms}}^2}{3R}.$$

Substituting the escape speed 11.2×10^3 m/s for v_{rms}, we have

$$T = \frac{(4 \times 10^{-3}\text{ kg/mol})(11.2 \times 10^3\text{ m/s})^2}{3(8.315\text{ J/mol·K})} = 2.01 \times 10^4\text{ K},$$

which is over 35 000 °F. This explains why the Earth is able to retain an atmosphere, while certain, much smaller, heavenly bodies cannot.

14.29 The molar mass of molecular hydrogen gas (H_2) is $M = 2.00$ g/mol $= 2.00 \times 10^{-3}$ kg/mol. The rms speed of the particles in a gas is

$$v_{\text{rms}} = \sqrt{\frac{3RT}{M}} = \sqrt{\frac{3(8.315\text{ J/mol·K})(300\text{ K})}{2.00 \times 10^{-3}\text{ kg/mol}}} = 1.93 \times 10^3\text{ m/s}.$$

Although this rms speed is considerably less than the escape speed, it is still close enough to it so that some hydrogen molecules in the distribution have speeds exceeding the escape speed and therefore escape. Heat transfer from the Earth's surface continually replenishes the supply of high speed molecules, so they too gradually escape. The same process also is involved in the phenomenon of evaporation.

14.33 The hydrogen atoms have a particle density of $\rho = 0.3$ particle/m^3, so the volume of one mole is

$$V = \frac{(1\text{ mol})(6.022 \times 10^{23}\text{ particle/mol})}{0.3\text{ particle/m}^3} = 2 \times 10^{24}\text{ m}^3.$$

Use the ideal gas law for one mole,

$$PV = nRT \implies P = \frac{nRT}{V} = \frac{(1.00\text{ mol})(8.315\text{ J/mol·K})(2.7\text{ K})}{2 \times 10^{24}\text{ m}^3} = 1 \times 10^{-23}\text{ Pa}.$$

14.37

a) Let m be the mass of the hydrogen atom and v its speed. Then the helium atom has mass $4m$. Let v' be its speed. Choose a coordinate system with $\hat{\mathbf{i}}$ parallel to $\vec{v}'$, and therefore antiparallel to $\vec{v}$. Since the initial kinetic energies are equal, we have

$$m\frac{v^2}{2} = 4m\frac{v'^2}{2} \implies v' = \frac{v}{2}.$$

Let v_{xa} and v'_{xa} be the x-components of the velocities of the hydrogen and helium atoms respectively, after the elastic collision. Since total momentum is conserved, and since $v' = \frac{v}{2}$,

$$4mv'\hat{\mathbf{i}} - mv\hat{\mathbf{i}} = 4mv'_{ax}\hat{\mathbf{i}} + mv_{ax}\hat{\mathbf{i}} \implies 2v\hat{\mathbf{i}} - v\hat{\mathbf{i}} = 4v'_{ax}\hat{\mathbf{i}} + v_{ax}\hat{\mathbf{i}}.$$

Hence,

(1) $$v_{ax} = v - 4v'_{ax}.$$

Since the collision is elastic, the total kinetic energy also is conserved. Hence

$$m\frac{v^2}{2} + 4m\frac{v'^2}{2} = m\frac{v_{ax}^2}{2} + 4m\frac{v'^2_{ax}}{2}.$$

Substitute $\dfrac{v}{2}$ for v' and simplify, to obtain

$$2v^2 = v_{ax}^2 + 4v'^{\,2}_{ax}.$$

Use equation (1) to substitute for v_{ax} and simplify again:

$$20v'^{\,2}_{ax} - 8vv'_{ax} - v^2 = 0 \text{ m}^2/\text{s}^2\,.$$

Factor to obtain

$$(10v'_{ax} + v)(2v'_{ax} - v) = 0 \text{ m}^2/\text{s}^2\,.$$

The solutions for v'_{ax} are

$$v'_{ax} = -\frac{v}{10} \quad \text{and} \quad v'_{ax} = \frac{v}{2}.$$

The second solution means the speed of the helium atom was unaffected, i.e., there was no collision, so the desired solution is

$$v'_{ax} = -\frac{v}{10}.$$

In this case, equation (1) implies

$$v_{ax} = \frac{7}{5}v.$$

The hydrogen atom increased its speed and kinetic energy while the helium atom decreased its speed and kinetic energy.

b)   If the two particles have momentum of equal magnitude before the collision, the total momentum before the collision is zero (the two particles are traveling in opposite directions). After the collision, the total momentum also is zero, so the particles merely reverse directions with unchanged speeds. Since the speeds are unchanged, the kinetic energy of each particle is unaffected by the collision.

14.41   The surface of the frying pan is two-dimensional, so the water drop can move in two spatial dimensions. Since it can't spin or otherwise distinguish itself, these are the only degrees of freedom that it has. Thus there are two degrees of freedom.

14.45   Helium is a monatomic gas. The internal energy U of a monatomic gas is given by

$$U = \frac{3nRT}{2} \implies n = \frac{2U}{3RT} = \frac{2(100 \text{ J})}{3(8.315 \text{ J/mol·K})(300 \text{ K})} = 2.67 \times 10^{-2} \text{ mol}.$$

14.49   Begin with the ideal gas law

$$PV = nRT.$$

The internal energy of a monatomic ideal gas is

$$U = \frac{3nRT}{2} \implies nRT = \frac{2U}{3}.$$

Substitute this expression for nRT into the ideal gas law:

$$PV = \frac{2U}{3}.$$

14.53

a)   The molar mass of molecular hydrogen gas is 2.00 g/mol, so the quantity of gas is 1.00 mol. The rms speed of the particles in the gas is

$$v_{rms} = \sqrt{\frac{3RT}{M}} = \sqrt{\frac{3(8.315 \text{ J/mol·K})(300 \text{ K})}{2.00 \times 10^{-3} \text{ kg/mol}}} = 1.93 \times 10^3 \text{ m/s}.$$

b)   Use the ideal gas law to find the pressure:

$$PV = nRT \implies P = \frac{nRT}{V} = \frac{(1.00 \text{ mol})(8.315 \text{ J/mol·K})(300 \text{ K})}{2.00 \times 10^{-3} \text{ m}^3} = 1.25 \times 10^6 \text{ Pa}.$$

c)   The temperature of the ideal gas does not change during a free expansion. Since the internal energy of an ideal gas is a function of temperature, the internal energy of the gas does not change.

d)   For an ideal gas at constant temperature, apply the ideal gas law to the initial and final states of the gas:

$$P_i V_i = nRT \quad \text{and} \quad P_f V_f = nRT \implies P_i V_i = P_f V_f$$

$$\implies P_f = \frac{P_i V_i}{V_f} = \frac{(1.25 \times 10^6 \text{ Pa})(2.00 \times 10^{-3} \text{ m}^3)}{4.00 \times 10^{-3} \text{ m}^3} = 6.25 \times 10^5 \text{ Pa}.$$

14.57

a)   The isochoric process is at constant volume and so is a vertical line in the P–V diagram. Its equation is $V = $ constant. An isothermal process is at constant temperature, and so follows an isotherm. Its equation is $PV = $ constant. An adiabatic expansion lowers the temperature, and so is below the isothermal process in the P–V diagram. Its equation is $PV^\gamma = $ constant. The three processes are shown in the P–V diagram below:

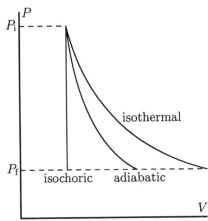

b)   The work done by the gas in each process is the area under the appropriate curve in the P–V diagram. There is no work done in the isochoric process since the area under its curve is zero.

c)   From the P–V diagram, the process with the greatest area under the curve is the isothermal process, so the gas does the most work during this process.

d)   There is no heat transfer to the gas during an adiabatic process.

    Consider the first law of thermodynamics, $Q = \Delta U + W$. During the isochoric process, there is no work done by the gas, so the absolute magnitude of the heat transfer to the gas is equal to the absolute magnitude of the change in its internal energy, $|Q_{\text{isochoric}}| = |\Delta U_{\text{isochoric}}|$. From Equation 14.21, the change in the internal energy is

$$|\Delta U_{\text{isochoric}}| = nc_V |\Delta T|.$$

From the ideal gas law at constant volume V_i,

$$|\Delta T| = \frac{V_i |\Delta P|}{nR} \implies |Q_{\text{isochoric}}| = nc_V \frac{V_i |\Delta P|}{nR} = c_V \frac{V_i |\Delta P|}{R}.$$

Use the ideal gas law again to substitute $\dfrac{nRT_i}{P_i}$ for V_i in the last expression:

(1)
$$|Q_{\text{isochoric}}| = c_V \frac{\dfrac{nRT_i}{P_i}|\Delta P|}{R} = nc_V T_i \frac{|\Delta P|}{P_i}.$$

For the isothermal process, there is no change in the internal energy, since the internal energy of an ideal gas is a function of the absolute temperature and that does not change in the isothermal process. Hence, the heat transfer to the gas during the isothermal process is equal to the work done by the gas:

$$|Q_{\text{isothermal}}| = |W_{\text{isothermal}}| = nRT_i \ln \frac{V_f}{V_i}.$$

Using the ideal gas law at constant temperature T_i, the ratio of the volumes can be written as a ratio of the pressures,

$$|Q_{\text{isothermal}}| = nRT_i \left| \ln \frac{P_i}{P_f} \right|.$$

Use this equation together with equation (1) to write

$$\frac{|Q_{\text{isothermal}}|}{|Q_{\text{isochoric}}|} = \frac{nRT_i \left| \ln \dfrac{P_i}{P_f} \right|}{nc_V T_i \dfrac{|\Delta P|}{P_i}} = \frac{RP_i}{c_V |\Delta P|} \left| \ln \frac{P_i}{P_f} \right|.$$

e)   Since the internal energy of an ideal gas is a function of its absolute temperature, the greatest magnitude change in the internal energy will be in the process that results in the greatest magnitude change in temperature. From the P–V diagram, it is the isochoric process that has the greatest absolute magnitude of change in the temperature.

14.61   For an isothermal process, we have

$$PV = \text{a constant.}$$

Differentiate both sides with respect to V using the product rule, and then solve for $\left(\dfrac{dP}{dV} \right)$.

(1)
$$\left(\frac{dP}{dV} \right) V + P \cdot 1 = 0 \implies \left(\frac{dP}{dV} \right)_T = -\frac{P}{V}.$$

For an adiabatic process, we have

$$PV^{\gamma} = \text{a constant.}$$

Differentiate both sides of this equation with respect to V using the product rule, and then solve for $\left(\dfrac{dP}{dV} \right)$.

(2)
$$\left(\frac{dP}{dV} \right) V^{\gamma} + P \cdot \gamma V^{\gamma-1} = 0 \implies \left(\frac{dP}{dV} \right)_{\text{adiabatic}} = -\gamma \frac{P}{V}.$$

Comparing equations (1) and (2) we see that

$$\left(\frac{dP}{dV} \right)_{\text{adiabatic}} = \gamma \left(\frac{dP}{dV} \right)_T.$$

Chapter 15

The Second Law of Thermodynamics

15.1 The maximum theoretical efficiency is that of a Carnot heat engine operating between the same two temperatures, so

$$\epsilon_{\text{Carnot}} = 1 - \frac{T_C}{T_H} = 1 - \frac{253 \text{ K}}{573 \text{ K}} = 0.558 = 55.8\%.$$

15.5 The maximum theoretical efficiency is that of a Carnot heat engine operating between the same two temperatures, so

$$\epsilon_{\text{Carnot}} = 1 - \frac{T_C}{T_H} = 1 - \frac{293 \text{ K}}{310 \text{ K}} = 0.055 = 5.5\%.$$

15.9 The efficiency of the Carnot heat engine is

$$\epsilon_{\text{Carnot}} = 1 - \frac{T_C}{T_H} = \frac{T_H - T_C}{T_H} \implies 0.200 = \frac{100 \text{ K}}{T_H} \implies T_H = 500 \text{ K} \quad \text{and} \quad T_C = T_H - 100 \text{ K} = 400 \text{ K}.$$

15.13

a) The efficiency of the heat engine is

$$\epsilon \equiv \frac{|W|}{|Q_H|}.$$

Consider a one second interval of time. During that interval

$$0.350 = \frac{100 \times 10^6 \text{ J}}{|Q_H|} \implies |Q_H| = 286 \times 10^6 \text{ J}.$$

From energy conservation, we have

$$|Q_H| = |Q_C| + |W| \implies 286 \times 10^6 \text{ J} = |Q_C| + 100 \times 10^6 \text{ J} \implies |Q_C| = 186 \times 10^6 \text{ J}.$$

Hence, the rate of heat transfer to the cooler environment is

$$186 \times 10^6 \frac{\text{J}}{\text{s}} = 186 \times 10^6 \text{ W} = 186 \text{ MW}.$$

b) Consider a one second interval of time. To warm the water the heat transfer to it is

$$Q_C = mc\Delta T \implies 186 \times 10^6 \text{ J} = m(4186 \text{ J/kg·K})(3.0 \text{ K}) \implies m = 1.5 \times 10^4 \text{ kg}.$$

Hence the *mass* of water needed each second is 1.5×10^4 kg. The mass is the density times the volume. The density of water is 1.00×10^3 kg/m^3. So the *volume* of water needed each second is

$$\frac{1.5 \times 10^4 \text{ kg}}{1.00 \times 10^3 \text{ kg/m}^3} = 15 \text{ m}^3.$$

Thus, the *flow* of water needed is 15 m^3/s.

15.17   Let the original temperatures of the reservoirs be T_H and T_C. The coefficient of performance of a Carnot refrigerator engine operating between these two temperatures is

$$K_{\text{Carnot}} = \frac{T_C}{T_H - T_C}.$$

Consider increasing the temperature of the cold reservoir by 10 K. Then the new coefficient of performance would be

$$K'_{\text{Carnot}} = \frac{T_C + 10\text{ K}}{T_H - (T_C + 10\text{ K})} = \frac{T_C + 10\text{ K}}{T_H - T_C - 10\text{ K}}.$$

Now consider decreasing the temperature of the hot reservoir by 10 K. The new coefficient of performance would be

$$K''_{\text{Carnot}} = \frac{T_C}{(T_H - 10\text{ K}) - T_C} = \frac{T_C}{T_H - T_C - 10\text{ K}}.$$

Note that the denominators for both K'_{Carnot} and K''_{Carnot} are the same, but the numerator of K'_{Carnot} is greater than the numerator of K''_{Carnot}. Hence, $K'_{\text{Carnot}} K''_{\text{Carnot}}$. It is better to increase the temperature of the cold reservoir by 10 K than to decrease the temperature of the hot reservoir by 10 K.

15.21

a)   The maximum theoretical efficiency is that of a Carnot engine operating between the same two temperatures:

$$\epsilon_{\text{Carnot}} = 1 - \frac{T_C}{T_H} = 1 - \frac{300\text{ K}}{500\text{ K}} = 0.400.$$

b)   The efficiency of the real engine is 0.250, and the work it does per cycle is 800 J. The efficiency is defined as

$$\epsilon \equiv \frac{|W|}{|Q_H|} \implies 0.250 = \frac{800\text{ J}}{|Q_H|} \implies |Q_H| = 3.20 \times 10^3 \text{ J}.$$

Energy conservation implies that

$$|Q_H| = |Q_C| + |W| \implies |Q_C| = |Q_H| - |W| = 3.20 \times 10^3 \text{ J} - 800\text{ J} = 2.40 \times 10^3 \text{ J}.$$

Since the heat transfer is *to* the cold reservoir, the answer is $Q_C = +2.40 \times 10^3$ J.

15.25

a)   The minimum number of joules exhausted to the room will be with the most efficient refrigerator engine, a Carnot refrigerator engine. You do not need to know the latent heat of vaporization because the calculation is done per joule of heat transfer. For a Carnot cycle, we have (from Equation 15.13)

$$\frac{|Q_C|}{|Q_H|} = \frac{T_C}{T_H} \implies |Q_H| = \frac{T_H}{T_C}|Q_C| = \frac{300\text{ K}}{77\text{ K}}(1.00\text{ J}) = 3.9\text{ J}.$$

Therefore, the heat transfer *to* the room is $Q_H = +3.9$ J.

b)   For a helium liquefier

$$|Q_H| = \frac{T_H}{T_C}|Q_C| = \frac{300\text{ K}}{4.2\text{ K}}(1.00\text{ J}) = 71\text{ J}.$$

The heat transfer *to* the room is +71 J.

15.29

a) The efficiency is defined as $\epsilon \equiv \dfrac{|W|}{|Q_H|}$. Energy conservation implies that

$$|Q_H| = |Q_C| + |W| \implies \epsilon = \frac{|W|}{|Q_C| + |W|} \implies 0.35 = \frac{|W|}{500 \text{ J} + |W|}$$

$$\implies 1.8 \times 10^2 \text{ J} + 0.35|W| = |W| \implies |W| = 2.8 \times 10^2 \text{ J}.$$

The work done by the heat engine is positive, so $W = +2.8 \times 10^2$ J.

b) Use energy conservation:

$$|Q_H| = |Q_C| + |W| = 500 \text{ J} + 2.8 \times 10^2 \text{ J} = 7.8 \times 10^2 \text{ J}.$$

The heat transfer is *to* the heat engine, and is therefore positive: $Q_H = +7.8 \times 10^2$ J.

15.33 The condensation occurs at constant temperature. Hence, the entropy change is

$$\Delta S = \int_i^f \frac{1}{T} \, dQ = \frac{1}{T} \int_i^f dQ = \frac{Q}{T},$$

where Q is the heat transfer "to" the steam necessary to condense it. (See example 15.6). Since the heat transfer really is from the steam when it condenses, Q is negative, so $Q = -mL_v$. Therefore

$$\Delta S = \frac{-mL_v}{T} = \frac{-(1.00 \text{ kg})(22.57 \times 10^5 \text{ J/kg})}{373.15 \text{ K}} = -6.05 \times 10^3 \text{ J/K}.$$

The negative entropy change does not violate the second law of thermodynamics because the kilogram of steam is *not* an isolated system during the condensation.

15.37

a) The ice warms to 0 °C, melts, then the melted water warms to the temperature of the ocean (which acts as a reservoir whose temperature is unchanged throughout the process). The entropy change of the ice is the sum of the changes associated with each step of the process:

$$\Delta S_{\text{ice}} = \Delta S_{\text{ice warming}} + \Delta S_{\text{ice melting}} + \Delta S_{\text{melted ice water warming}}$$

$$= m_{\text{ice}} c_{\text{ice}} \ln \frac{T_f}{T_i} + \frac{m_{\text{ice}} L_f}{273.15 \text{ K}} + m_{\text{ice melted}} c_{\text{water}} \ln \frac{T_f}{T_i}$$

$$= (0.100 \text{ kg})(2050 \text{ J/kg·K}) \ln \frac{273.15 \text{ K}}{268.15 \text{ K}} + \frac{(0.100 \text{ kg})(3.335 \times 10^5 \text{ J/kg})}{273.15 \text{ K}}$$

$$+ (0.100 \text{ kg})(4186 \text{ J/kg·K}) \ln \frac{283.15 \text{ K}}{273.15 \text{ K}}$$

$$= 3.79 \text{ J/K} + 122 \text{ J/K} + 15.1 \text{ J/K}$$

$$= 141 \text{ J/K}.$$

b) The entropy change of the ocean is $\Delta S = \dfrac{-Q}{T_{\text{ocean}}}$. We are told that $T_{\text{ocean}} = 283.15$ K, so to find ΔS we need Q, the heat transfer needed to warm the ice, melt it, and then warm it up to the ocean temperature. Thus,

$$Q = Q_{\text{to warm ice}} + Q_{\text{to melt ice}} + Q_{\text{to warm melted ice water}}$$

$$= m_{\text{ice}} c_{\text{ice}} \Delta T_{\text{ice}} + m_{\text{ice}} L_f + m_{\text{melted ice}} c_{\text{water}} \Delta T_{\text{melted water}}$$

$$= (0.100 \text{ kg})(2050 \text{ J/kg·K})(5.00 \text{ K}) + (0.100 \text{ kg})(3.335 \times 10^5 \text{ J/kg}) + (0.100 \text{ kg})(4186 \text{ J/kg·K})(10.00 \text{ K})$$

$$= 1.03 \times 10^3 \text{ J} + 3.34 \times 10^4 \text{ J} + 4.19 \times 10^3 \text{ J}$$

$$= 3.86 \times 10^4 \text{ J}.$$

Hence the entropy change of the ocean is

$$\Delta S_{\text{ocean}} = \frac{-Q}{T_{\text{ocean}}} = \frac{-3.86 \times 10^4 \text{ J}}{283.15 \text{ K}} = -136 \text{ J/K}.$$

c)   The entropy change of the ice-ocean system is the sum of the entropy changes of the ice and ocean,

$$\Delta S_{\text{total}} = \Delta S_{\text{ice}} + \Delta S_{\text{ocean}} = 141 \text{ J/K} + (-136 \text{ J/K}) = 5 \text{ J/K}.$$

15.41

a)   Since a Carnot cycle consists of alternating adiabatic (constant S) and isothermal (constant T) processes, the shape of a carnot cycle on a T–S diagram is a rectangle:

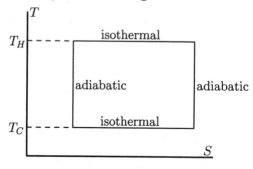

b)   The definition of the differential entropy change is

$$dS \equiv \frac{dQ}{T} \implies dQ = T \, dS.$$

The heat transfer to the system during each process is the area under the curve in the T–S diagram for that particular process. Hence, the area enclosed by the cycle on the T-S diagram is the total heat transfer to the system during one cycle.

c)   The efficiency of a Carnot engine is given by

$$\epsilon_{\text{Carnot}} = 1 - \frac{T_C}{T_H} = \frac{T_H - T_C}{T_H}.$$

The numerator and denominator of this expression for the efficiency may be interpreted as the distances indicated on the T–S diagram of the Carnot cycle shown below, and thus, ϵ_{Carnot} may be interpreted as the ratio of those distances:

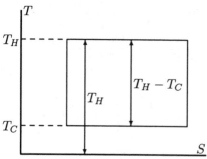

15.45

a)   The efficiency of the heat engine is

$$\epsilon = \frac{|W|}{Q_H}.$$

Consider the interval Δt during which the engine completes one cycle. During this time the work done by the engine is $W = (1.00 \text{ kW})\Delta t$, so

$$0.250 = \frac{(1.00 \text{ kW})\Delta t}{|Q_H|} \implies \frac{|Q_H|}{\Delta t} = 4.00 \text{ kW}.$$

Hence the *rate* of heat transfer (the heat flow) from the high temperature reservoir to the engine is 4.00 kW.

We may also compute the rate of heat transfer from the engine to the cold reservoir. During the interval Δt, the heat transfer to the cold reservoir is

$$|Q_C| = |Q_H| - |W| = (4.00 \text{ kW})\Delta t - (1.00 \text{ kW})\Delta t = (3.00 \text{ kW})\Delta t. \implies \frac{|Q_C|}{\Delta t} = 3.00 \text{ kW}.$$

So, the heat flow from the heat engine to the cold reservoir is 3.00 kW.

b) Consider the system to be the two reservoirs and the engine. The total entropy change in one cycle is

$$\Delta S_{\text{total}} = \Delta S_{\text{hot reservoir}} + \Delta S_{\text{cold reservoir}} + \Delta S_{\text{engine}}.$$

The entropy change of the engine is 0 J/K because it returns to the same state of thermodynamic equilibrium after each cycle. The temperatures of the reservoirs are unaffected by the heat transfer to or from them. Thus, during one cycle, we have

$$\begin{aligned} \Delta S_{\text{total}} &= \frac{Q_{\text{hot reservoir}}}{T_H} + \frac{Q_{\text{cold reservoir}}}{T_C} + 0 \text{ J/K}. \\ &= \frac{(-4.00 \text{ kW})\Delta t}{1000 \text{ K}} + \frac{(3.00 \text{ kW})\Delta t}{400 \text{ K}} \\ &= (3.50 \text{ W/K})\Delta t. \end{aligned}$$

Therefore, the time rate at which the entropy of the entire system increases is $\dfrac{\Delta S}{\Delta t} = 3.50 \text{ W/K}$. (Use the negative sign for Q_H because we are computing the entropy change for the hot reservoir, and heat is flowing *from* the hot reservoir).

15.49 You and the environment act as reservoirs, transferring heat with no change in temperature. The entropy change of the 'you+environment' system is

$$\Delta S = \Delta S_{\text{you}} + \Delta S_{\text{environment}}.$$

The heat transfer is from you to the environment. Your body temperature is about $37 \text{ °C} = 310 \text{ K}$. Consider a length of time Δt. During this time, the entropy increase is

$$\begin{aligned} \Delta S = \Delta S_{\text{you}} + \Delta S_{\text{environment}} &= \frac{-(100 \text{ W})\Delta t}{310 \text{ K}} + \frac{(100 \text{ W})\Delta t}{293 \text{ K}} \\ &= (-0.323 \text{ W/K})\Delta t + (0.341 \text{ W/K})\Delta t = (0.018 \text{ W/K})\Delta t. \end{aligned}$$

There are 8.6400×10^4 seconds in one day. Hence the entropy of the system increases by

$$(0.018 \text{ W/K})\Delta t = (0.018 \text{ W/K})(8.6400 \times 10^4 \text{ s}) = 1.6 \times 10^3 \text{ J/K}$$

each day.

15.53

a) During an adiabatic process there is no heat transfer to or from the system, so

$$Q = 0 \text{ J}.$$

b)   Since $Q = 0$ J in an adiabatic process, the first law of thermodynamics says

$$0 \text{ J} = Q = \Delta U + W \implies \Delta U = -W = -2000 \text{ J}.$$

c)   The entropy change is

$$\Delta S = \int_i^f \frac{dQ}{T}.$$

But, since the process is adiabatic, $dQ = 0$, and therefore $\Delta S = 0$ J/K.

d)   Since the change in the internal energy of the ideal gas is negative, and the internal energy is proportional to the absolute temperature ($\Delta U = nc_V \Delta T$), the temperature of the gas decreased.

e)   The change in the internal energy of an ideal gas is

$$\Delta U = nc_V \Delta T \implies \Delta T = \frac{\Delta U}{nc_V}.$$

For a diatomic gas, $c_V = \dfrac{5}{2}R$, so

$$\Delta T = \frac{2\Delta U}{5nR} = \frac{2(-2000 \text{ J})}{5(2.00 \text{ mol})(8.315 \text{ J/mol·K})} = -48.1 \text{ K}.$$

15.57   When the containers are brought into thermal contact, the heat transfer to the cooler one is all from the warmer one, so the total heat transfer to the two considered as one system is zero. The heat transfer between the two containers occurs at constant pressure. The final temperature T of the combined system is found from

$$nc_P \Delta T_{\text{cooler gas}} + nc_P \Delta T_{\text{warmer gas}} = 0 \text{ J} \implies \Delta T_{\text{cooler gas}} + \Delta T_{\text{warmer gas}} = 0 \text{ K}$$

$$\implies T - T_1 + T - T_2 = 0 \text{ K} \implies T = \frac{T_1 + T_2}{2}.$$

(Which you might have guessed before doing the algebra!)

The entropy change is

$$\Delta S_1 + \Delta S_2 = \int_{T_1}^T \frac{nc_P}{T}\, dT + \int_{T_2}^T \frac{nc_P}{T}\, dT$$

$$= nc_P \left(\int_{T_1}^T \frac{dT}{T} + \int_{T_2}^T \frac{dT}{T} \right)$$

$$= nc_P (\ln T - \ln T_1 + \ln T - \ln T_2)$$

$$= nc_P \ln \frac{T^2}{T_1 T_2}$$

$$= nc_P \ln \frac{\left(\dfrac{T_1 + T_2}{2}\right)^2}{T_1 T_2}$$

$$= nc_P \ln \frac{(T_1 + T_2)^2}{4T_1 T_2}.$$

In order to show that the entropy change $\Delta S \geq 0$, we must show that $\ln \dfrac{(T_1 + T_2)^2}{4T_1 T_2} \geq 0$. This is equivalent to showing that $\dfrac{(T_1 + T_2)^2}{4T_1 T_2} \geq 1$, which in turn is equivalent to showing that

(1) $$(T_1 + T_2)^2 \geq 4T_1 T_2.$$

However,

$$(T_1 + T_2)^2 \geq 4T_1T_2 \quad \Leftrightarrow \quad T_1^2 + 2T_1T_2 + T_2^2 \geq 4T_1T_2 \quad \Leftrightarrow \quad T_1^2 - 2T_1T_2 + T_2^2 \geq 0 \quad \Leftrightarrow \quad (T_1 - T_2)^2 \geq 0.$$

The last inequality is certainly true, since the square of *any* number is greater than or equal to zero. Therefore (1) is true. Thus, the entropy change of the isolated system is greater than or equal to zero, in conformity with the second law of thermodynamics.

15.61 Shake the box many times and find the experimental probability of securing each macrostate after a shake. If, after many shakes, you never see a result with 0 heads, you should become suspicious. The following table describes a *true* 4-penny system. The table lists the possible macrostates, the number of microstates for each macrostate, and the probability of each macrostate. There are a total of $2^4 = 16$ possible microstates, all equally likely.

macrostate (# of heads)	# of microstates	probability of macrostate
0	$\frac{4!}{0!\,4!} = 1$	$\frac{1}{16} = 0.0625$
1	$\frac{4!}{1!\,3!} = 4$	$\frac{4}{16} = 0.2500$
2	$\frac{4!}{2!\,2!} = 6$	$\frac{6}{16} = 0.3750$
3	$\frac{4!}{3!\,1!} = 4$	$\frac{4}{16} = 0.2500$
4	$\frac{4!}{4!\,0!} = 1$	$\frac{1}{16} = 0.0625$

With one two-headed penny in a 4-penny system, the system cannot have a macrostate with 0 heads. There are therefore a total of only $2^3 = 8$ possible microstates, all equally likely. The macrostates and microstates are tabulated below.

macrostate (# of heads)	microstates	probability of macrostate
1	HTTT	$\frac{1}{8} = 0.125$
2	HHTT HTHT HTTH	$\frac{3}{8} = 0.375$
3	HHHT HHTH HTHH	$\frac{3}{8} = 0.375$
4	HHHH	$\frac{1}{8} = 0.125$

15.65

a)   The left- and right-sides of the box are like heads and tails of pennies. The number of microstates for each macrostate is

$$\frac{N!}{n!\,(N-n)!},$$

where N is the total number of particles in the system, in our case 6, and n is the number on, say, the left side. The macrostates and number of corresponding microstates are tabulated below:

macrostate #on left-side	# microstates Ω
0	$\dfrac{6!}{0!\,6!} = 1$
1	$\dfrac{6!}{1!\,5!} = 6$
2	$\dfrac{6!}{2!\,4!} = 15$
3	$\dfrac{6!}{3!\,3!} = 20$
4	$\dfrac{6!}{4!\,2!} = 15$
5	$\dfrac{6!}{5!\,1!} = 6$
6	$\dfrac{6!}{6!\,0!} = 1$

b)   There is only 1 microstate corresponding to the macrostate with all 6 particles on the left side of the box. The entropy of this macrostate is

$$S = k \ln \Omega = k \ln 1 = 0 \text{ J/K}.$$

c)   The entropy of the most probable macrostate is

$$S = k \ln 20 = (1.381 \times 10^{-23} \text{ J/K}) \ln 20 = 4.137 \times 10^{-23} \text{ J/K}$$

15.69

a)   The total number N of pennies in the system is 20, so the total number of microstates is $2^N = 2^{20} = 1\,048\,576$. The number of microstates associates with macrostate n, where n is the number of heads, is

$$\Omega = \frac{N!}{n!\,(N-n)!}.$$

Use this formula to determine Ω and $\dfrac{S}{k} = \ln \Omega$.

n	Ω	$\frac{S}{k} = \ln\Omega$
0	$\frac{20!}{0!\,20!} = 1$	0
1	$\frac{20!}{1!\,19!} = 20$	3.00
2	$\frac{20!}{2!\,18!} = 190$	5.25
3	$\frac{20!}{3!\,17!} = 1140$	7.04
4	$\frac{20!}{4!\,16!} = 4845$	8.49
5	$\frac{20!}{5!\,15!} = 15\,504$	9.65
6	$\frac{20!}{6!\,14!} = 38\,760$	10.57
7	$\frac{20!}{7!\,13!} = 77\,520$	11.26
8	$\frac{20!}{8!\,12!} = 125\,970$	11.74
9	$\frac{20!}{9!\,11!} = 167\,960$	12.03
10	$\frac{20!}{10!\,10!} = 184\,756$	12.13

n	Ω	$\frac{S}{k} = \ln\Omega$
11	$\frac{20!}{11!\,9!} = 167\,960$	12.03
12	$\frac{20!}{12!\,8!} = 125\,970$	11.74
13	$\frac{20!}{13!\,7!} = 77\,520$	11.26
14	$\frac{20!}{14!\,6!} = 38\,760$	10.57
15	$\frac{20!}{15!\,5!} = 15\,504$	9.65
16	$\frac{20!}{16!\,4!} = 4845$	8.49
17	$\frac{20!}{17!\,3!} = 1140$	7.04
18	$\frac{20!}{18!\,2!} = 190$	5.25
19	$\frac{20!}{19!\,1!} = 20$	3.00
20	$\frac{20!}{20!\,0!} = 1$	0

b) See part c) below.

c) Here are graphs of Ω vs. n and $\frac{S}{k}$ vs. n.

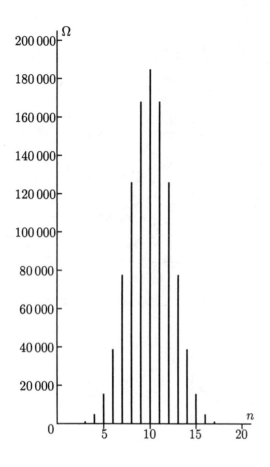

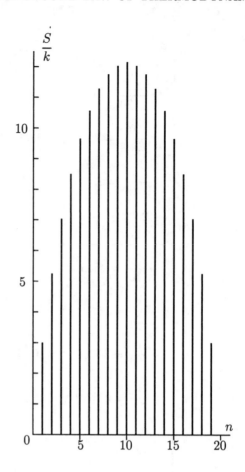